T0032915

101 Things I Learned in Engineering School

Other books in the 101 Things I Learned® series

101 Things I Learned in Advertising School

101 Things I Learned in Architecture School (MIT Press)

101 Things I Learned in Business School

101 Things I Learned in Culinary School

101 Things I Learned in Fashion School

101 Things I Learned in Film School

101 Things I Learned in Law School

101 Things I Learned in Urban Design School

101 Things I Learned® in Engineering School

John Kuprenas with Matthew Frederick

THREE RIVERS PRESS
NEW YORK

Published in the United States by Three Rivers Press,
an imprint of the Crown Publishing Group,
a division of Penguin Random House LLC, New York.
crownpublishing.com

Originally published in the United States in slightly different form by
Grand Central Publishing, a division of Hachette Book Group, in 2013.

Library of Congress Cataloging-in-Publication Data is available upon request.

ISBN 978-1-5247-6196-7
Ebook ISBN 978-1-5247-6197-4

Printed in China

Illustrations by Matthew Frederick
Cover illustration by Matthew Frederick

10 9 8 7 6 5 4

From John

To my family

Author's Note

Engineers view their profession as fascinating, creative, and full of interesting challenges. Those outside it often regard it as repetitive, mechanical, and frustrating.

What is evident from both perspectives is that engineering is complex. It requires intensive study in mathematics, physics, and chemistry, which fills most of the first two years of the college curriculum. While focusing on these concerns, the curriculum tends to provide very little context. When I was a beginning engineering student, I was frustrated that the calculations and abstract concepts I was learning in the classroom were difficult to tie to the real world. The engineering curriculum presented a lot of trees, and very little forest.

101 Things I Learned in Engineering School flips this around. It introduces engineering largely through its context, by emphasizing the common sense behind some of its fundamental concepts, the themes intertwined among its many specialties, and the simple abstract principles that can be derived from real-world circumstance. It presents, I believe, some clear glimpses of the forest as well as the trees within it.

It is my hope that this book will interest and enlighten college students seeking context for their developing mathematical and scientific knowledge, inspire practicing engineers to reflect on the subtle relationships in their field, and encourage the layperson to see the engineering world as engineers do: fascinating, creative, challenging, collaborative, and unfailingly rewarding.

John Kuprenas

Acknowledgments

From John

Thanks to Weston Hester, Keith Crandall, Ben Gerwick, William C. Ibbs, Povindar K. Mehta, David Blackwell, the inspiration of the books at Skylight and Powell's, and the conversations at Figaro Café.

From Matt

Thanks to Tricia Boczkowski, Regina Brooks, Nancy Byrnes, Sorche Fairbank, Venkataramana Gadhamshetty, Harmonie Hawley, Matt Inman, Andrea Lau, Dave McNeilly, Amanda Patten, Angeline Rodriguez, Aaron Santos, Simon Schelling, Molly Stern, and Rick Wolff. Special thanks to Marshall Audin, Myev Bodenhofer, and David Mallard for their ideas, help, and support.

101 Things I Learned in Engineering School

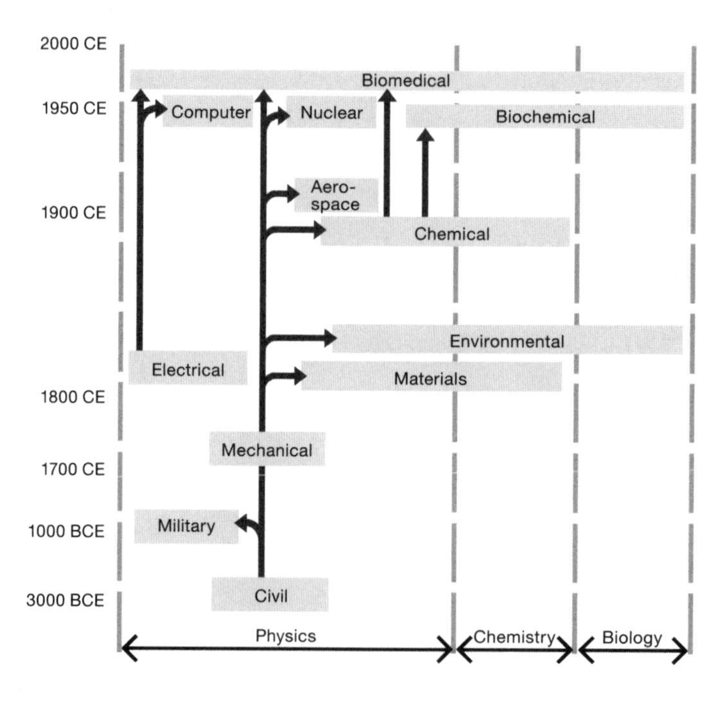

Engineering family tree

Civil engineering is the grandparent of all engineering.

In its early days during the Roman Empire, civil engineering was synonymous with military engineering. Their kinship was still strong when the first engineering school in America was founded in 1802 at the U.S. Military Academy at West Point, New York. USMA graduates planned, designed, and supervised the construction of much of the nation's early infrastructure, including roads, railways, bridges, and harbors, and mapped much of the American West.

input

output

2

Engineering succeeds and fails because of the black box.

Engineering is a field of specialties, with different individuals or teams working on different aspects of a project. A **black box** conceptually contains the knowledge and processes of an engineering specialty. On multidisciplinary design teams, the output from one discipline's black box becomes the input for the black boxes of one or more other disciplines. The designer of a fuel system, for example, works within a "fuel system black box" that produces an output for the engine designer; the engine designer's black box outputs to the automatic transmission designer, and so on.

Design solutions aren't created linearly, however; teams work in complex, interconnected ways. Hence, the black box model works best when employed as a momentary ideal that is adjusted and redefined throughout the design process as constraints become evident, opportunities emerge, prototypes are tested, and goals are clarified. It fails when expected to be permanent and orderly.

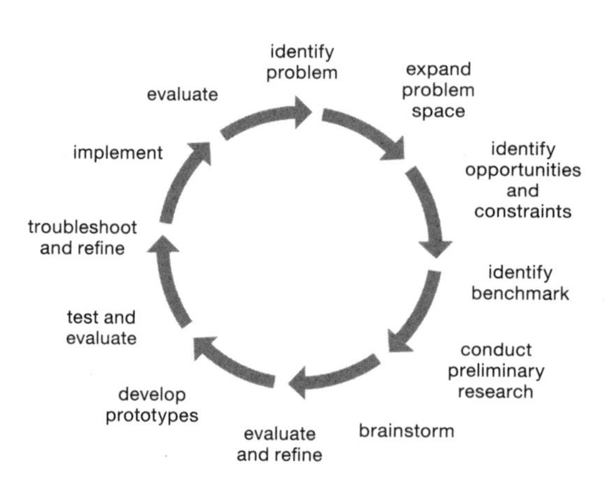
identify problem
expand problem space
identify opportunities and constraints
identify benchmark
conduct preliminary research
brainstorm
evaluate and refine
develop prototypes
test and evaluate
troubleshoot and refine
implement
evaluate

3

The heart of engineering isn't calculation; it's problem solving.

School may teach the numbers first, but calculation is neither the front end of engineering nor its end goal. Calculation is one means among many to finding a solution that provides useful, objectively measurable improvement.

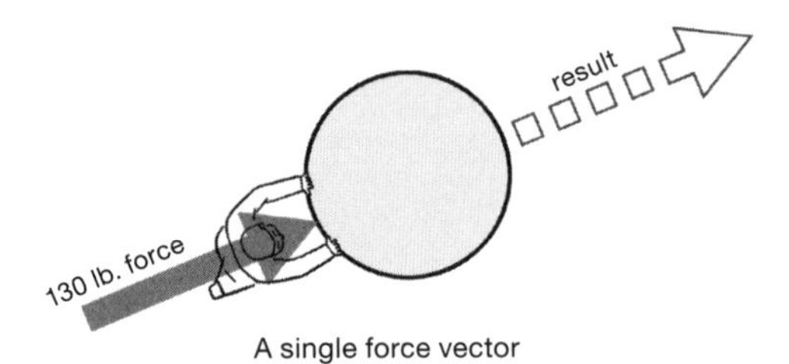

A single force vector

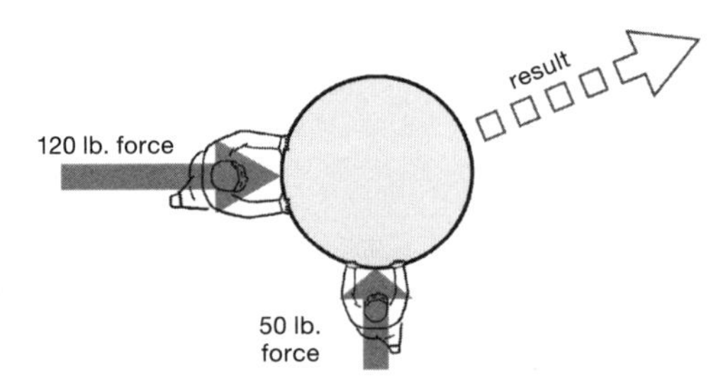

Two equivalent force vectors

You are a vector.

A force is expressed graphically by a vector. A vector's length represents its magnitude, and its direction is given in relation to the *x, y,* and *z* axes. Every person has a gravity force vector with a magnitude (weight) measured in pounds or newtons, and a direction toward the center of the earth. Any single vector can be replaced by more than one component vectors, and vice versa, as long as they yield an equivalent net result.

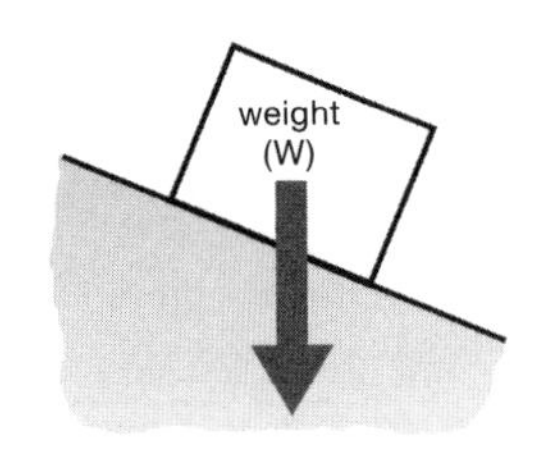

Block on a ramp

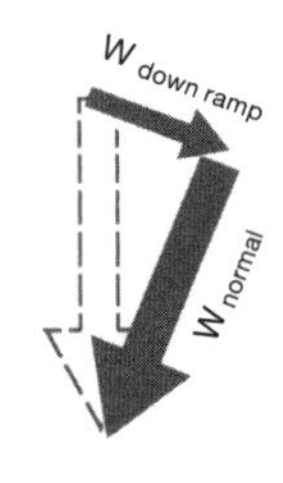

Gravity component vectors

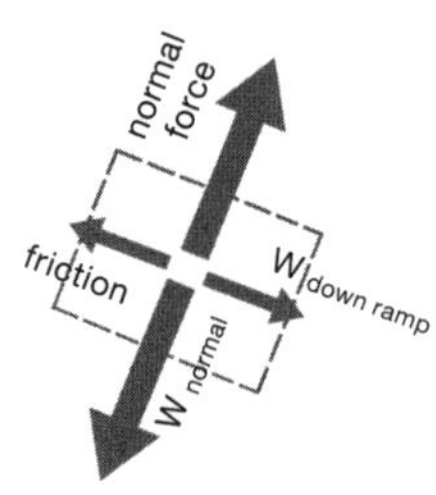

Free body diagram

5

Every problem is built on familiar principles.

Every problem has embedded in it a “hook”—a familiar, elemental concept of statics, physics, or mathematics. When overwhelmed by a complex problem, identify those aspects of it that can be grasped with familiar principles and tools. This may be done either intuitively or methodically, as long as the tools you ultimately use to solve the problem are scientifically sound. Working from the familiar will either point down the path to a solution, or it will suggest the new tools and understandings that need to be developed.

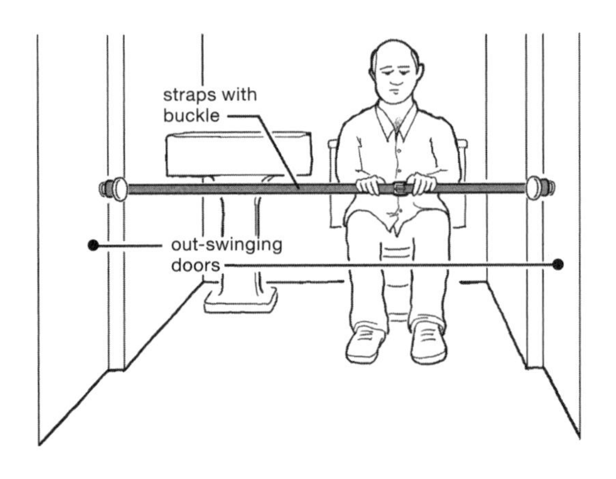

How the former Hotel Louis XIV in Quebec prevented guests
from locking each other out of the shared bathrooms

6

Every problem is unique.

Engineering problem solving relies on the familiar, but invention is also called for. Some problem-solving tools are developed through rote and repetition, some emerge intuitively, some rote-learned tools become intuitive over time, and some come out of necessity and even desperation. Add the tools you develop while solving each problem to your toolbox, to use on future problems. More important, add to your toolbox the methods by which you *discovered* the new tools.

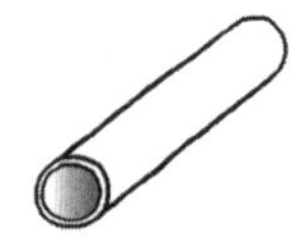

Straight pipe
friction loss of 5.5'
per 100' of run

90° elbow
friction loss equal to
4.0' straight run

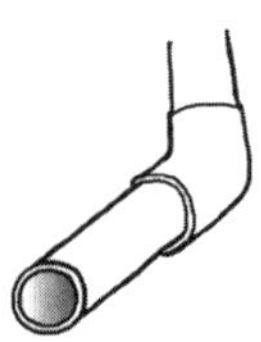

45° elbow
friction loss equal to
2.0' straight run

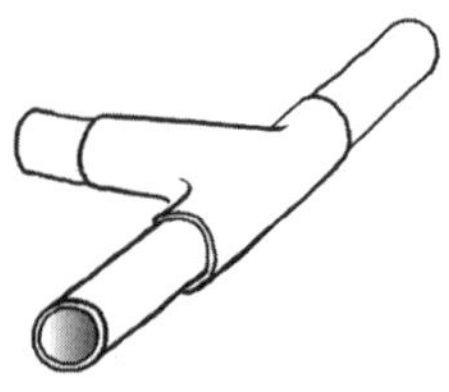

Tee, side outlet
friction loss equal to
8.0' straight run

To simplify pressure loss calculations,
convert all components to an equivalent length of straight pipe.
(Assumes 1½" dia. PVC pipe at 30 gallons per minute initial flow)

7

"Inside every large problem is a small problem struggling to get out."

—TONY HOARE

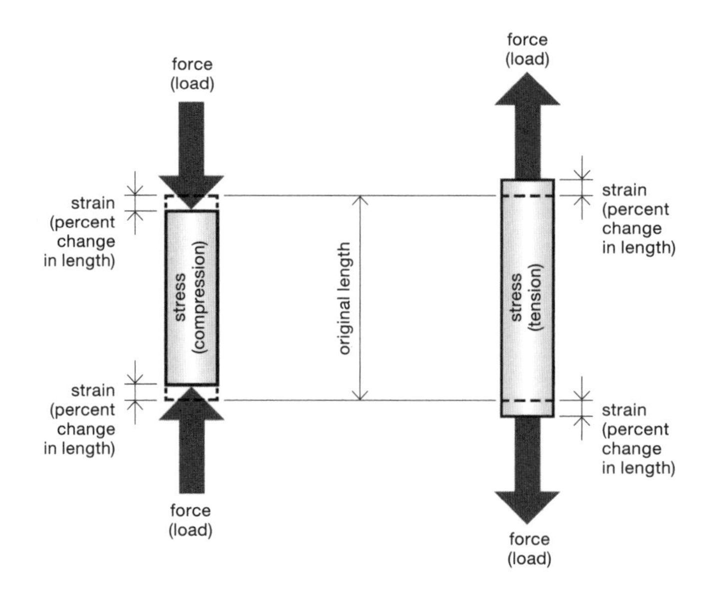
force
(load)
strain
(percent
change
in length)
stress
(compression)
original length
stress
(tension)
strain
(percent
change
in length)
force
(load)

An object receives a force, experiences stress, and exhibits strain.

The words "force," "stress," and "strain" are used somewhat interchangeably in the lay world, and may even be used with less than ideal rigor by engineers. However, they have different meanings.

A **force,** often called a "load," exists external to and acts upon a body, and can cause it to change speed, direction, or shape. Examples of forces include water pressure on a submarine hull, snow loads on a bridge, and wind loads on the sides of a skyscraper.

Stress is the "experience" of a body—its internal resistance to an external force acting on it. Stress is force divided by area, and is expressed in units such as pounds per square inch.

Strain is a product of stress. It is the measurable percentage of deformation or change in an object, such as an increase or decrease in length.

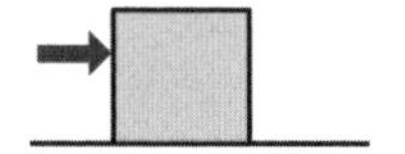

Object remains stationary

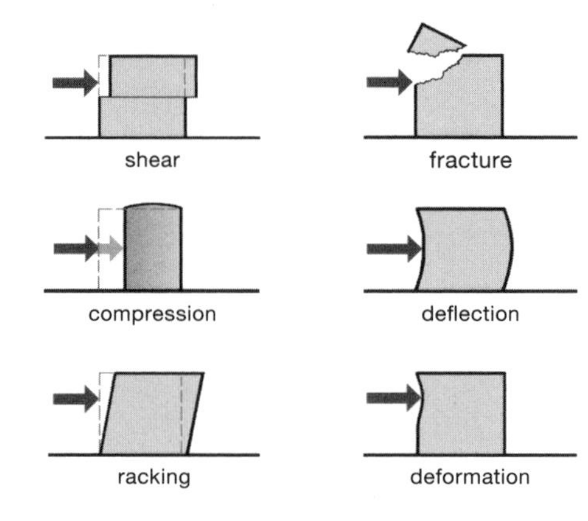

Object changes shape

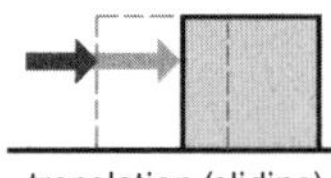

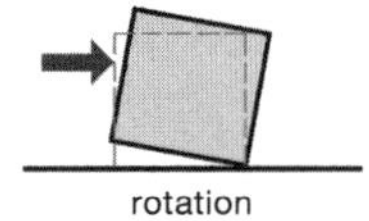

Object moves

9

When a force acts on an object, three things can happen.

An object that receives a force will remain stationary, move, or change shape—or undergo a combination of these reactions. Mechanical engineering generally seeks to exploit movement, while structural engineering seeks to prevent or minimize it. Most engineering disciplines aim to minimize changes in the shape of a designed object.

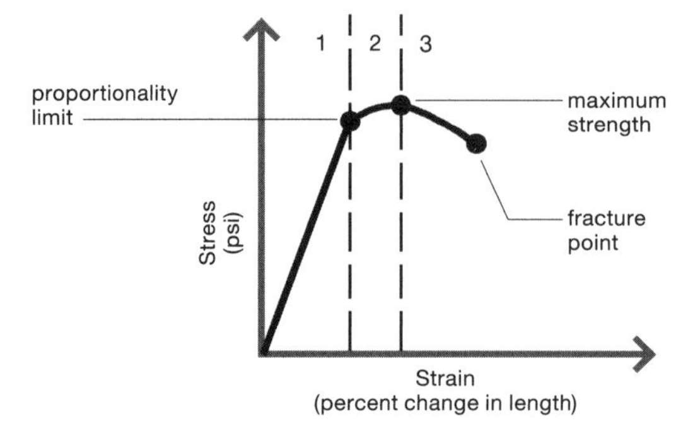

Simplified stress-strain curve

As a force acting on a fixed object increases, three things happen.

1. **Proportional elongation phase:** When an object, such as a steel bar, is subjected to a stretching (tensile) force, it initially will deform in proportion to the loads placed on it. If load *x* causes the bar to deform *d,* 2*x* will cause deformation 2*d,* 3*x* will cause 3*d,* and so on. If the load is removed, the bar will return to its original length.

2. **Disproportional elongation phase:** Beyond a certain point of loading, which varies among materials, an object will deform at a rate greater than the rate of increase in loading. If load 10*x* causes deformation 10*d,* load 10.5*x* may cause 20*d.* When the load is removed, the material will not quite return to its original length.

3. **Ductility phase:** If loading is further increased, the material will become visibly deformed and will soon fracture.

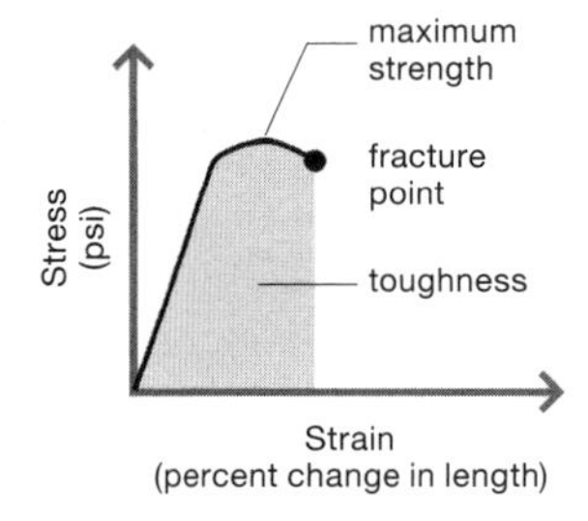

Stiff, strong, but brittle
(low-ductility) material

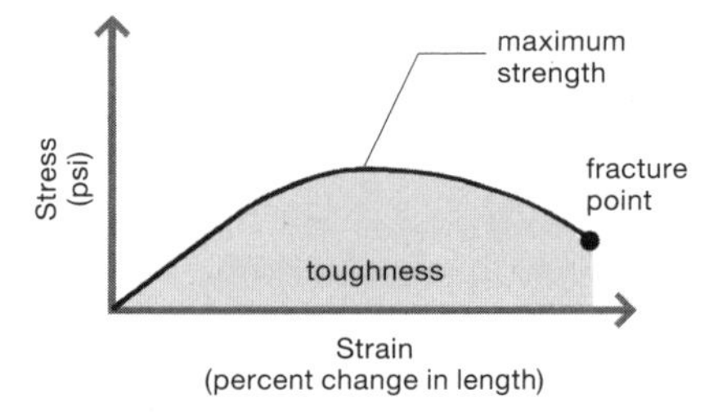

Less stiff, less strong, high ductile,
ultimately tougher material

Simplified stress-strain curves

Four material characteristics

Stiffness/elasticity concerns the lengthening or shortening of a material under loading. Stiffness is resistance to change in length; elasticity is the ability to return to original size and shape. Stiffness is measured formally by the **modulus of elasticity,** which is the slope of the straight line portion of the stress-strain curve: the steeper, the stiffer.

Strength is a measure of a material's ability to accept a load. The maximum strength of a material (usually tested in tension rather than compression) is represented by the highest point on the stress-strain curve.

Ductility/brittleness is the extent a material deforms or elongates before fracturing. A highly ductile material is taffy-like, and its stress-strain curve extends far to the right. A very brittle material is chalk-like; its curve ends abruptly after reaching maximum strength.

Toughness is an overall measure of a material's ability to absorb energy before fracture. It is represented by the total area under the stress-strain curve.

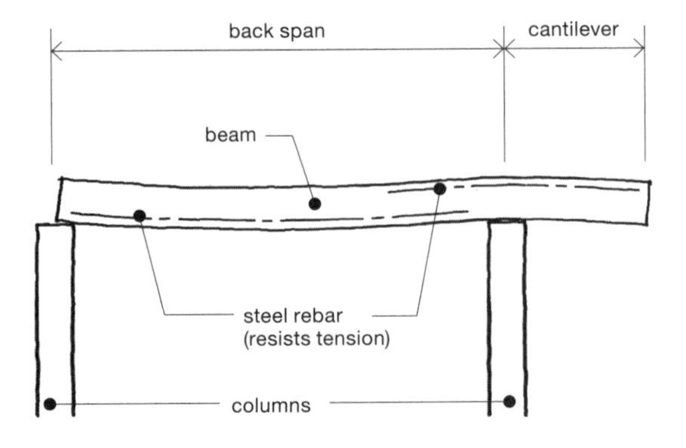

Steel-reinforced concrete beam

Materials compete.

Materials shrink and grow with fluctuations in atmospheric conditions, and change in strength, shape, size, and elasticity as they age. Where different materials have complementary properties, they can be intricately combined. Steel and concrete expand and contract at near-identical rates with changes in temperature; if they did not, a steel-reinforced concrete beam would tear itself apart upon ordinary temperature fluctuation.

More often, materials are not neutral toward each other. They compete for electrons where in contact, inducing corrosion; their sizes and shapes change at different rates upon variations in temperature, humidity, and air pressure; and they respond variably to wear, tear, and maintenance. An airplane tire and the wheel on which it is mounted, for example, will react in different ways to the rapid changes in temperature, air pressure, and loading they will undergo in ordinary use. The integrity of their relationship will be maintained only if the system is designed to perform across the entire range of its components' physical behaviors.

Anode
(more active)

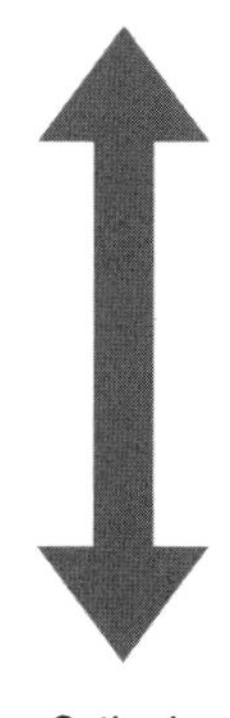

magnesium
zinc
aluminum
steel or iron
lead
nickel
brass
copper
bronze
stanless steel 304
Monel metal
silver
gold
platinum

Cathode
(less active)

Partial galvanic series

13

A battery works because of corrosion.

On the surface of all metals are loosely bound electrons. When two metals are placed in contact, the atoms of each compete to attract the electrons. The more "noble" metal (the **cathode**) attracts electrons from the more "active" metal (**anode**). The movement of electrons causes the anode to corrode, and produces an electric current. A common household battery generates current with a carbon-zinc cell, in which the zinc corrodes preferentially to the carbon (often considered a "metalloid," which behaves similarly to a metal).

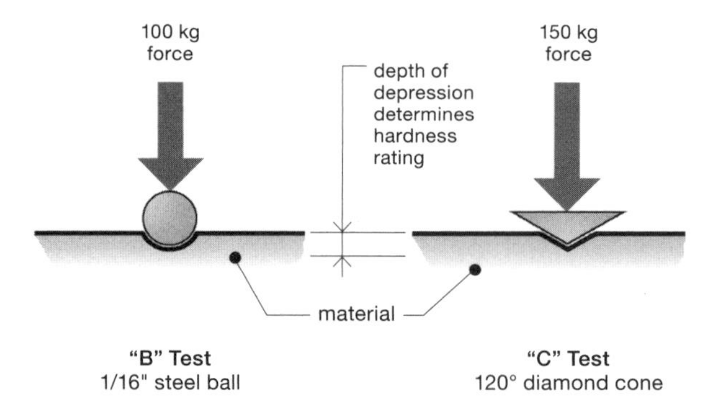

Rockwell hardness test

14

Harder materials don't ensure longevity.

In 1915, a ship was built with a hull of Monel, a relatively new and very hard alloy of nickel, copper, and iron. Expectations were that the *Sea Call* would have an exceptionally long life, as Monel is extremely corrosion resistant and is excellent for wet applications. Unfortunately, the 214' long, 34' wide vessel had to be scrapped after six weeks of use. The Monel hull was fully intact, but the Monel had caused the ship's steel frame and fasteners to deteriorate beyond use from electrolytic interaction in the saltwater environment.

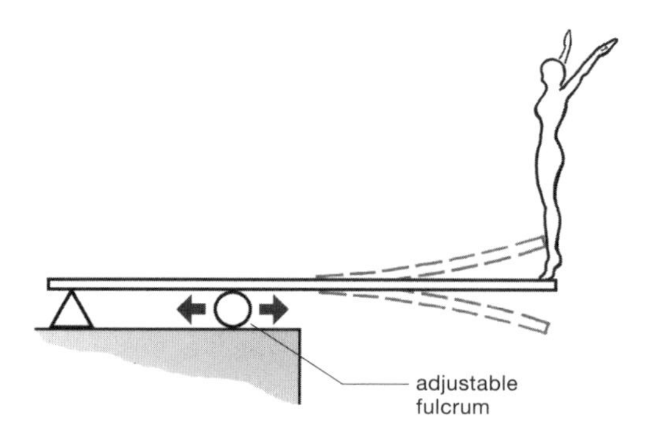

With each bounce, a diver stores energy in the board. By coordinating each landing with the board's natural frequency, the height of the takeoff is increased.

Soldiers shouldn't march across a bridge.

15

A structural member vibrates in response to normal loads and impacts, in the manner of a plucked guitar string. The **natural** or **resonant frequency** of an object is the time it takes to complete one cycle of movement (fully back and forth or up and down) upon disturbance.

When a force acts repeatedly on a structural member, and at a rate that matches its natural frequency, the member's response is enhanced with every cycle. The effects range from loud humming (such as when vibrations from a building's mechanical equipment coincide with a beam's natural frequency) to uncomfortable oscillation to occasional collapse. Many relatively small earthquakes have induced significant damage when their wave frequency has matched that of affected buildings. In 2000, thousands of pedestrians celebrating the opening of the London Millennium Footbridge inadvertently induced oscillation when their walking rhythms matched the structure's natural frequency. As they swayed in response to the unanticipated movement, they inadvertently increased it. The bridge was closed following the event and the structural system was repaired.

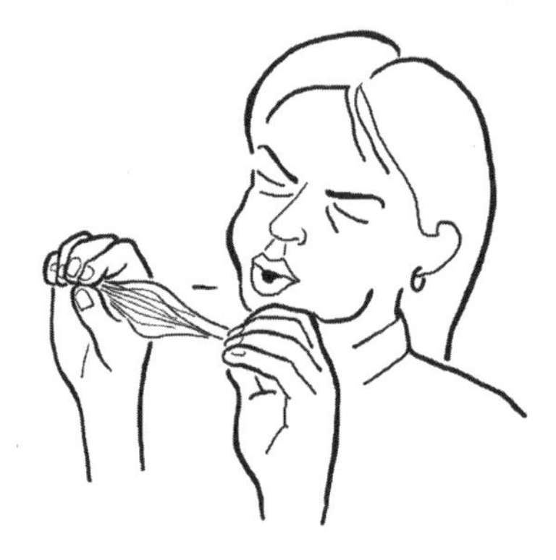

Aeroelastic flutter in a paper strip

Why Galloping Gertie collapsed

The Tacoma Narrows suspension bridge in Washington State was the third longest of its kind when completed. Despite unusual levels of motion during construction, "Galloping Gertie" opened to the public 1940. In November, it began to heave violently as a lone driver, Leonard Coatsworth, drove across it with his cocker spaniel. Unable to continue, and unable to remove "Tubby" from the car, Coatsworth fled on foot. Several unsuccessful attempts were made to rescue Tubby, but eventually the dog, car, and bridge fell into Puget Sound.

The Washington State Highway Department determined that the collapse was not the result of rhythmic wind gusts "exciting" the structure's natural resonance, as is often argued. Rather, the disaster began with **aeroelastic flutter** (a vibrational response to air movement), leading to **torsional flutter** (repetitive twisting). The 2,800' long, 39' wide main span was inherently wind-vulnerable, as its main girders were made of solid steel plates that were only 8' deep. By comparison, an earlier proposal for the bridge used a 25' deep, open-web stiffening truss.

Ten years after the collapse, a replacement bridge was built. "Sturdy Gertie" incorporated the original approach ramps and main piers, but used a 33' deep stiffening truss.

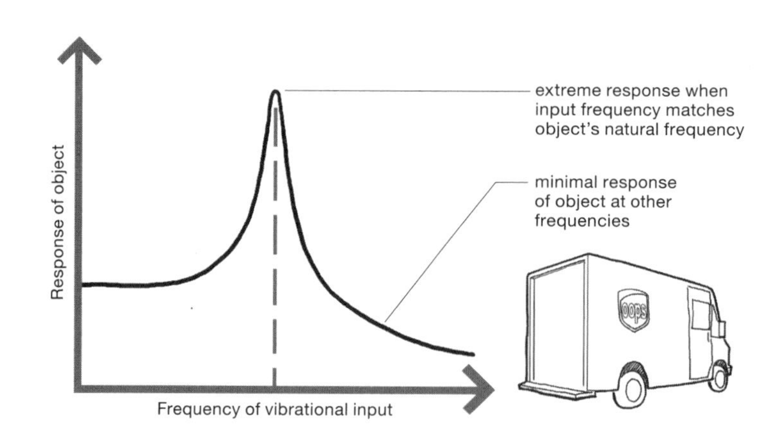
Response of object
Frequency of vibrational input
extreme response when input frequency matches object's natural frequency
minimal response of object at other frequencies
oops

17

Softer materials aren't always more protective.

Packaging engineers have found that very few packages are accidentally dropped during transport to the user, and far fewer are dropped from a height that will damage the contents. But while impact damage during transit is rare, every product is subject to vibrational input from the vehicles in which it is shipped. The wrong type of cushioning material can amplify vehicular vibration and cause a sensitive item to fail if the net vibration imparted to the item happens to be at its natural frequency. An improperly designed package thus can destroy the product it is meant to protect.

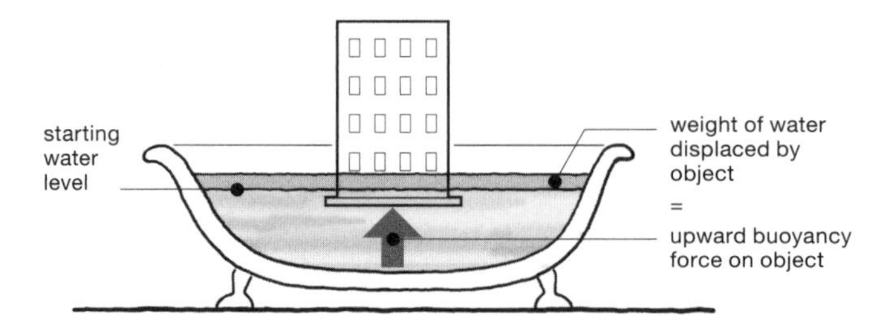

Archimedes' principle

18

Buildings want to float.

Buoyant uplift force on an object equals the weight of the water that the object displaces. If a building's lower floors extend below the groundwater level, buoyancy will seek to lift it—even if the displaced water is dispersed throughout the soil. Flotation is unlikely in a completed building, but a deep foundation, basement, or underground parking garage will seek to float to the surface of the earth without the weight of the building pressing down on it. For this reason, an underground storage tank must be attached to a concrete mass at least equal to the weight of the groundwater the tank will potentially displace.

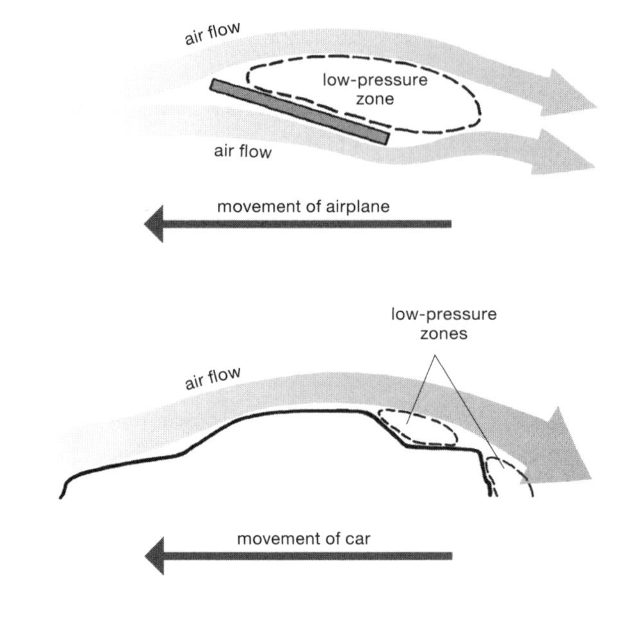

Simplified air flow diagrams

Automobiles want to fly.

An airplane wing in cross section is shaped as an airfoil, somewhat in the manner of a warped, tapered lozenge. But an airplane can fly even with a completely flat wing set at an angle. As the plane moves forward, a low-pressure zone is produced immediately above the wing, "sucking" the plane into the sky. However, an airfoil shape has far less drag and functions much more efficiently.

Low pressure zones are similarly created through an automobile's forward motion. A vacuum zone typically results behind a moving car, somewhat impeding its forward progress. If the car has a traditional sedan shape, a low-pressure area can also be created above the trunk, producing rear-end lift. Above about 70 mph, the driver's control can be noticeably affected. Approaching 200 mph, a car may become airborne.

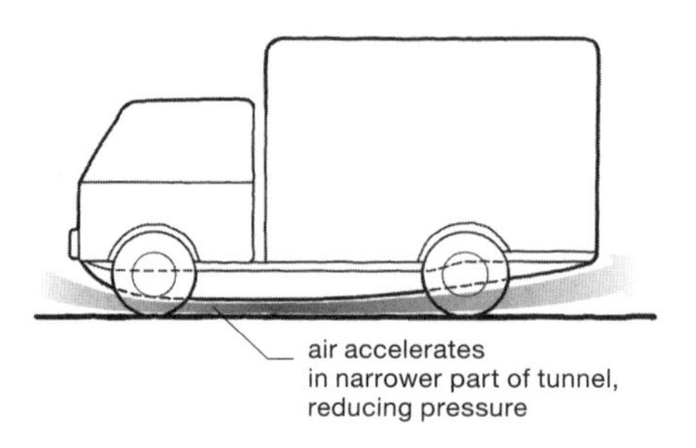

After a drawing from U.S. patent 4,386,801 by Colin Chapman et al.

The ground effect

The rear wing on a race car counteracts lift by introducing downforce at the rear of the vehicle. However, it also increases drag, reducing aerodynamic efficiency.

British inventor Colin Chapman sought a more efficient alternative. On the underside of his racing vehicles, he created a front-to-rear air channel, the top of which was shaped similarly to an inverted airfoil. Combined with very low ground clearance and side skirts, air moving under the vehicle was directed through a narrower zone, which caused the air to accelerate. As faster-moving air naturally has lower pressure, the vehicle was "sucked" to the road. The "ground effect" proved so effective when Chapman's Lotus team introduced it into Formula One racing that it was quickly banned.

There was a trade-off to Chapman's device: if a ground-effect car traveling at high speed was bumped, the air channel could be disrupted, resulting in catastrophic loss of control. But the genius of Chapman's invention is undeniable, for he inverted the perplexing problem of creating more "push-down": he worked from the opposite side and created more "pull-down."

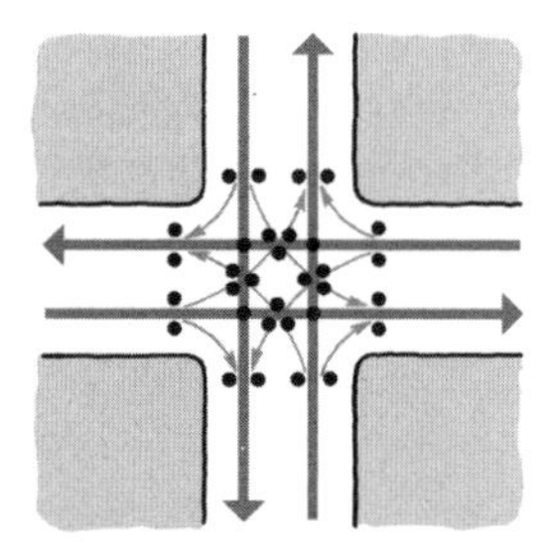

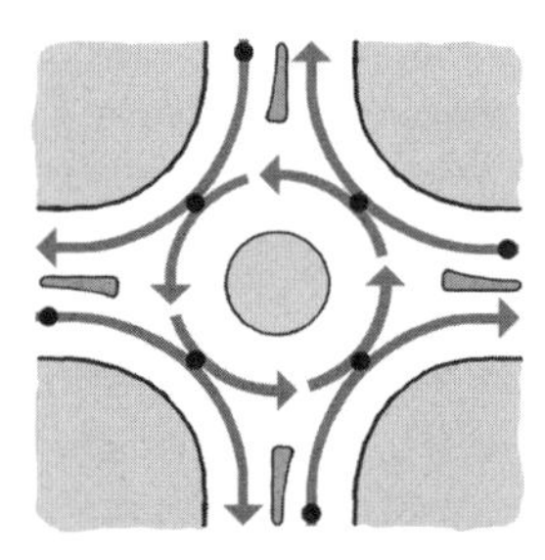

	Conventional intersection	Roundabout intersection
Vehicle conflict points (•)	32	8
Lane capacity per hour	1300—1500	1800
Operating speed	up to 55mph	15—25mph
Angle of collision	up to 90°	low/glancing

A roundabout is the safest, most efficient intersection.

Where roundabouts have replaced conventional intersections, traffic delays have been reduced up to 89%, accidents 37 to 80%, injuries 30 to 75%, and fatalities 50 to 70%. Up to an eightfold return on investment due to accident reductions has been indicated.

A civil engineering team at the University of Texas found that crossings with flashing lights are the most dangerous intersection type, with an accident rate approximately 3 times greater than stoplight intersections and 5 to 6 times greater than roundabouts.

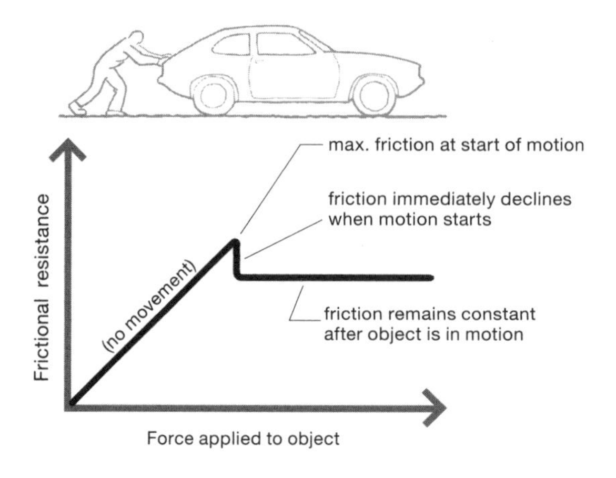
max. friction at start of motion
friction immediately declines when motion starts
friction remains constant after object is in motion
(no movement)
Frictional resistance
Force applied to object

22

Friction is the enemy of a rolling object, but it is what allows it to roll.

An object is slowed by friction between it and the surface on which it slides or rolls, as microscopic bumps and ridges on each catch on to those of the other. The greater the friction, the more a wheel's efficiency is reduced and the more heat is produced. The lesser the friction, the more freely and efficiently the wheel rolls. This suggests that a state of zero friction would allow a wheel to roll with perfect efficiency. But the wheel would not roll at all, because of an absence of traction; instead, it would slide.

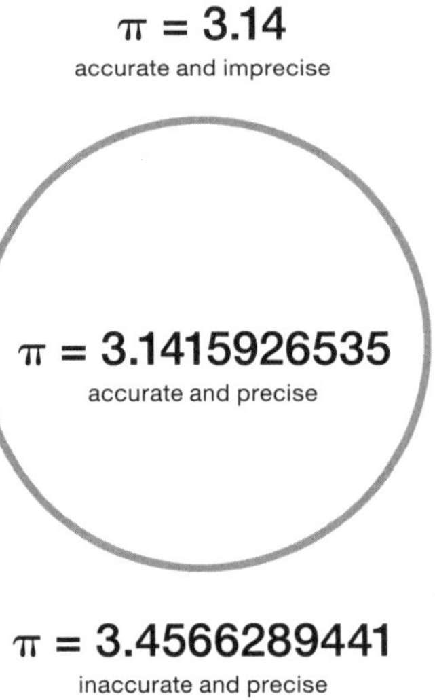

$$\pi = 3.14$$

accurate and imprecise

$$\pi = 3.1415926535$$

accurate and precise

$$\pi = 3.4566289441$$

inaccurate and precise

23

Accuracy and precision are different things.

Accuracy is the absence of error; **precision** is the level of detail. Effective problem solving requires always being accurate, but being only as precise as is helpful at a given stage of problem solving. Early in the problem-solving process, accurate but imprecise methods, rather than very exact methods, will facilitate design explorations while minimizing the tracking of needlessly detailed data.

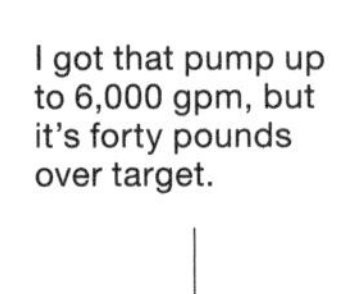
I got that pump up to 6,000 gpm, but it's forty pounds over target.

We're at our limit. See if someone else on the team can give you forty pounds.

24

There's always a trade-off.

Lightness versus strength, response time versus noise, quality versus cost, responsive handling versus soft ride, speed of measurement versus accuracy of measurement, design time versus design quality . . . it is impossible to maximize the response to every design consideration. Good design is not maximization of every response, or even compromise among them; it's optimization among alternatives.

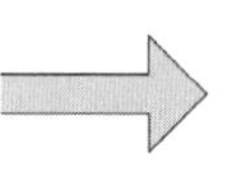
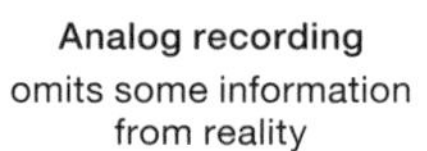

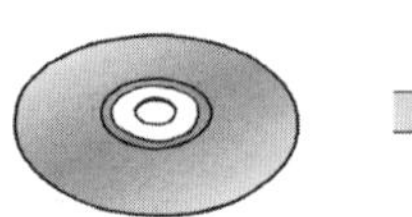
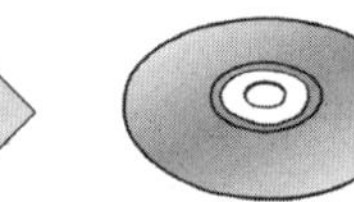
Analog recording
omits some information from reality
Digital copy
omits some information from analog recording
Reality
Digital recording
omits some information from reality
Digital copy
100% accurate to digital recording

25

Quantification is approximation.

Engineering follows the laws of science, but nature does not. As a system of understanding created by humans, science is contained within reality. Nature follows itself; science is our remarkable but imperfect attempt to explain it. Quantification is exact not unto reality, but unto itself.

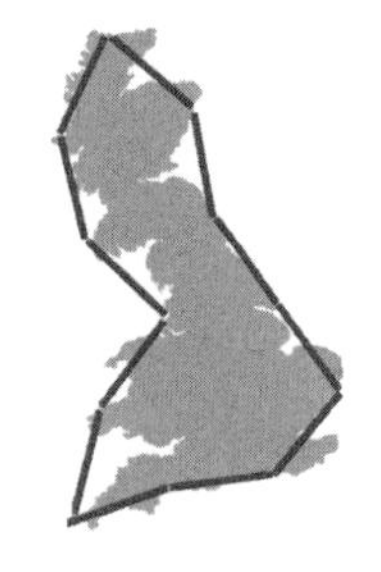

200 km measuring unit
coastline = 2,400 km

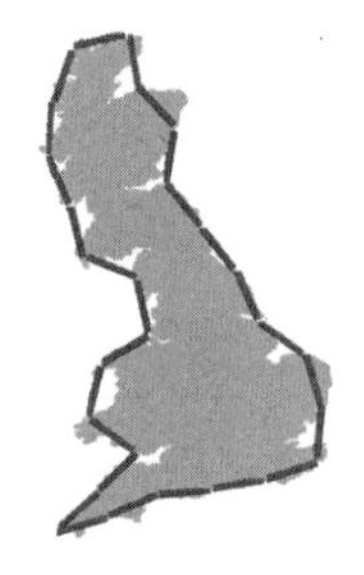

100 km measuring unit
coastline = 2,600 km

50 km measuring unit
coastline = 3,100 km

UK Ordnance Survey
coastline = 17,820 km

As a measuring device gets more precise, the measurement of the perimeter of an irregular object approaches infinity.

26

Random hypothesis #1

You don't fully understand something until you quantify it. But you understand nothing at all if all you do is quantify.

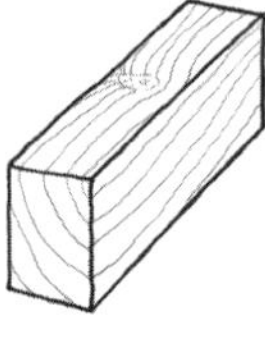
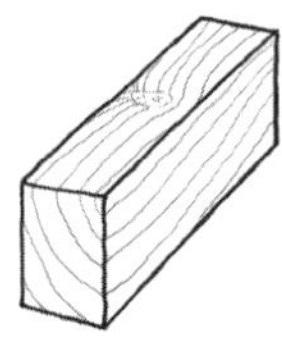
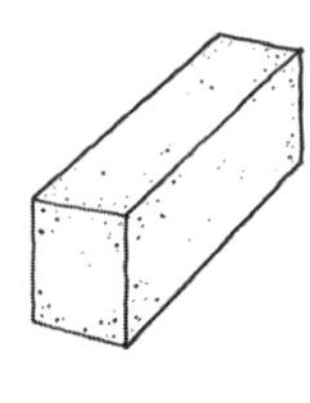
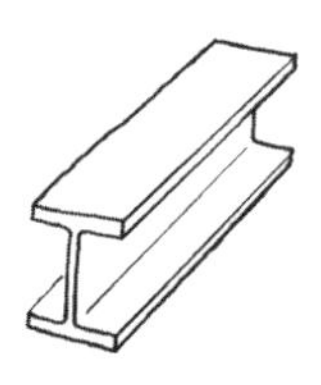

	Douglas fir	**Concrete**	**A36 Steel**
Maximum strength determined in laboratory testing	7,430 psi (compression)	4,000 psi (compression)	50,000 psi (tension)
Design strength used in calculations	1,350 psi	3,000 psi	36,000 psi
Approximate safety margin	**5.5**	**1.3**	**1.4**

Engineers wear a belt and suspenders.

All construction materials are laboratory tested to determine their structural properties, such as the amount they stretch and compress under loading and the maximum load they will accept before failing. The test results lead to the designation of a formal **design strength** for the material, which engineers subsequently use in real-world structural calculations. However, design strength is always established at a point lower than that at which the material failed, to allow for variability in quality.

Manufactured materials such as concrete and steel are of comparatively uniform quality, and variability from piece to piece is comparatively small. A wood beam, however, might have come from a diseased tree, have reacted atypically to drying, or contain an unusual number of knots. Consequently, the design strength for wood is much lower than that determined in laboratory testing.

Engineers commonly build in additional safety margins by overestimating loads, rounding their calculations to the conservative side, and selecting a structural member larger or thicker than the one their calculations call for.

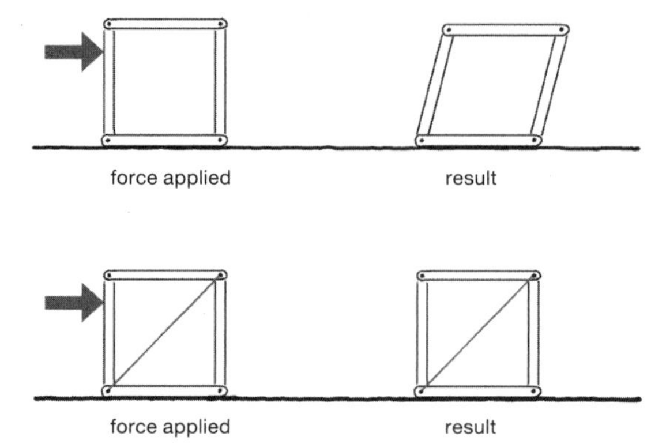
force applied
result
force applied
result

28

A triangle is inherently stable.

A triangle differs from other linear shapes in that its sides and angles are interdependent: a change cannot be made to an angle without altering the length of at least one side, and vice versa. By comparison, a square can be deformed into a parallelogram without changing a side.

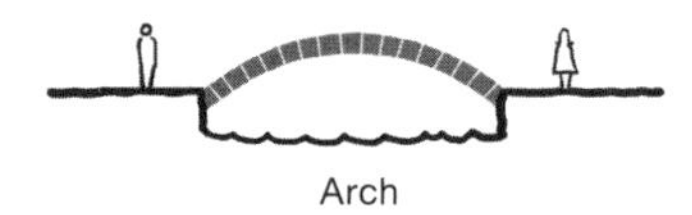

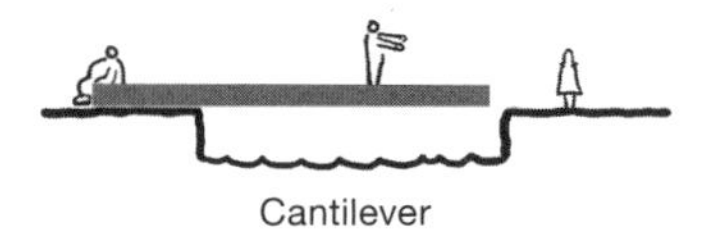

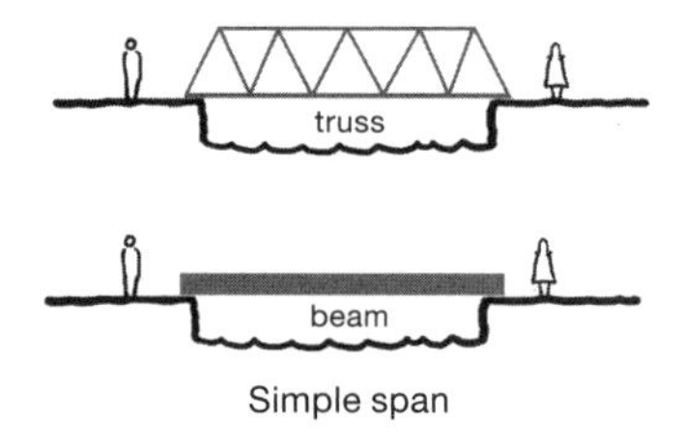

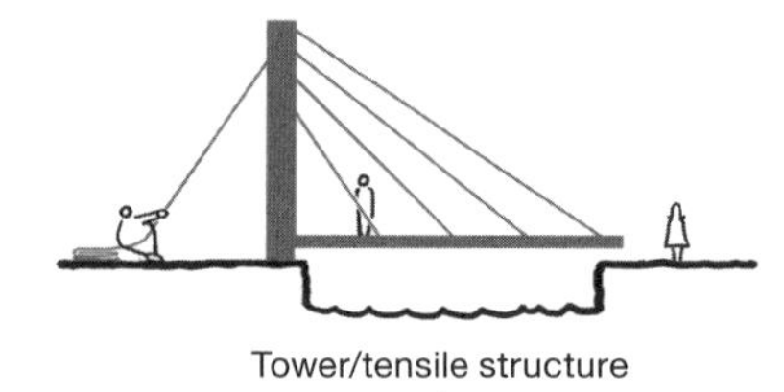

Four ways to span

29

The complexity of a truss is a product of simplicity.

A truss is a complex form of beam that takes advantage of the inherent stability of the triangle. By starting with a triangle and adding two legs at a time, a series of interdependent triangles form a stable structure capable of spanning long distances, using a fraction of the material used by an ordinary beam.

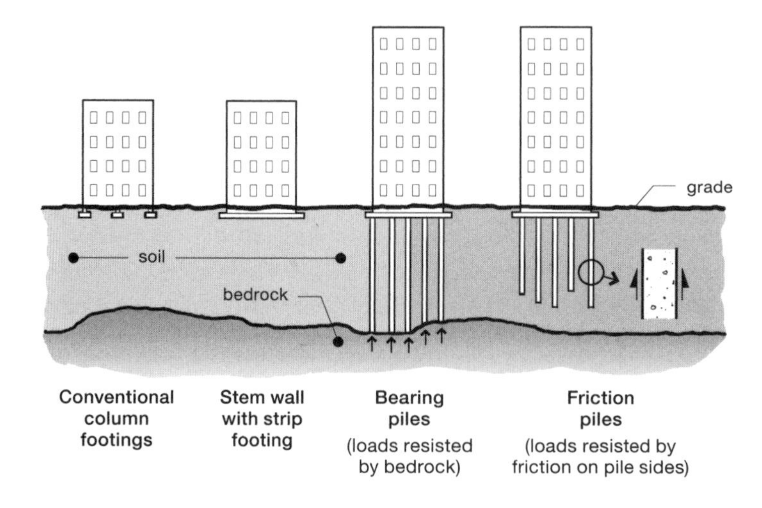

Common foundation/footing types

Structures are built from the bottom up, but designed from the top down.

The lower structural elements of a building support elements above them. Before a given structural element can be designed, the elements above it must be designed. But structural design cannot be performed in one top-to-bottom pass. Make numerous passes of increasing rigor and precision, from schematic through final design, before deciding how to best transfer loads downward to the earth.

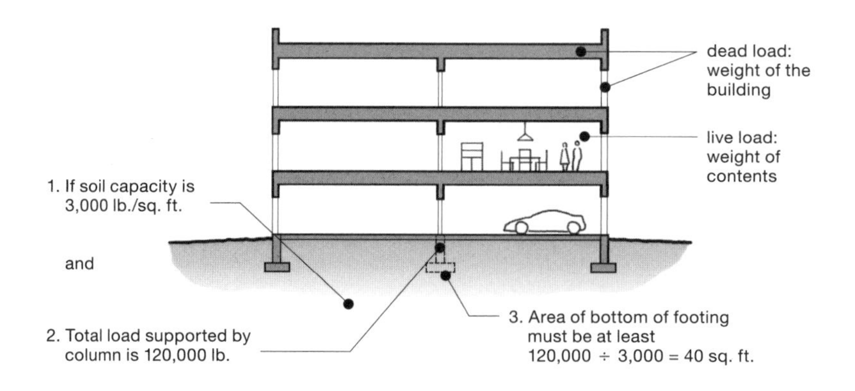
dead load:
weight of the
building
live load:
weight of
contents
1. If soil capacity is
3,000 lb./sq. ft.
and
2. Total load supported by
column is 120,000 lb.
3. Area of bottom of footing
must be at least
120,000 ÷ 3,000 = 40 sq. ft.

The contents of a building might weigh more than the building.

Dead load is the weight of a building itself and is almost entirely constant over the life of the building. It includes the structure (beams, columns, joists, etc.); primary building systems (exterior wall, windows, roofing, interior finishes, etc.), permanent architectural elements (stairways, partitions, flooring materials, etc.), and mechanical systems (heating, cooling, plumbing, electrical, etc.).

Live loads change over the life of a building. They come from people, furniture, vehicles, wind, earthquakes, snow, impacts of foreign objects, and similarly variable sources.

Total load is transmitted to the building foundation, then to the earth. The load received by a given area of a footing cannot exceed the bearing capacity of the soil, or the footing will sink.

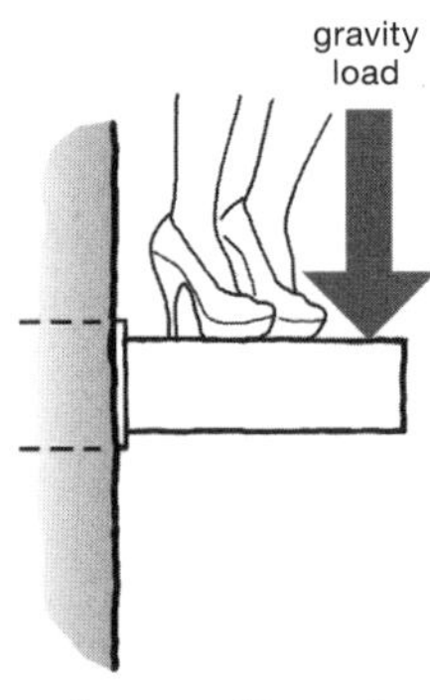

Beam cantilevered
from a wall

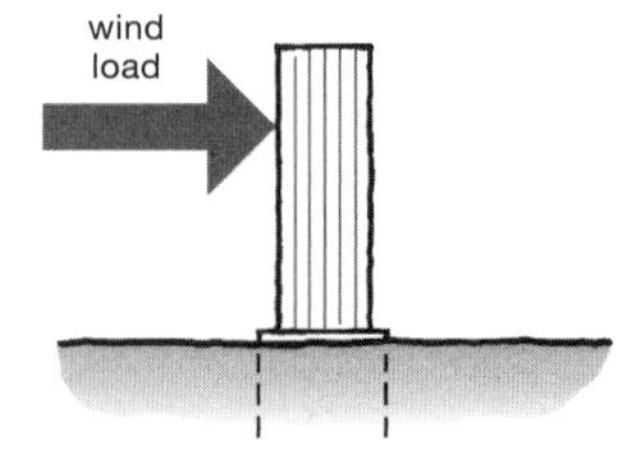

Skyscraper "cantilevered"
from the earth

32

A skyscraper is a vertically cantilevered beam.

The primary structural design challenge of a skyscraper is not resolving vertical (gravity) loads, but resisting lateral loads from wind and earthquakes. For this reason, tall structures function and are designed conceptually as large beams cantilevered from the ground.

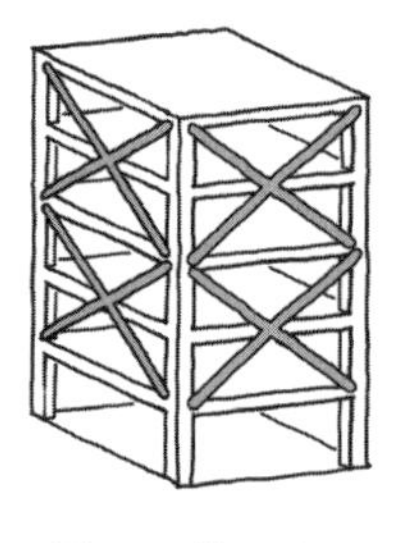

Diagonal bracing
triangulation among structural members resists lateral forces

Shear wall
extra-stiff construction resists lateral forces in the direction of its surface

Floor diaphragm
extra-stiff construction resists lateral forces in the plane of the floor

Three ways of increasing lateral rigidity

Earthquake design: let it move a lot or not at all.

Earthquakes are typically characterized by lateral (side to side) movement. A structure can resist this force by being either very flexible or very rigid. In a **flexible structure,** beam-column connections rotate with relative freedom when stressed, with perhaps some damping or diagonal shock absorbers. A **rigid structure** relies on very strong connections among structural members, with isolators (essentially, large rubber doughnuts) at the building base. In both systems, seismic energy is damped so that building inhabitants experience a fraction of the earthquake's forces.

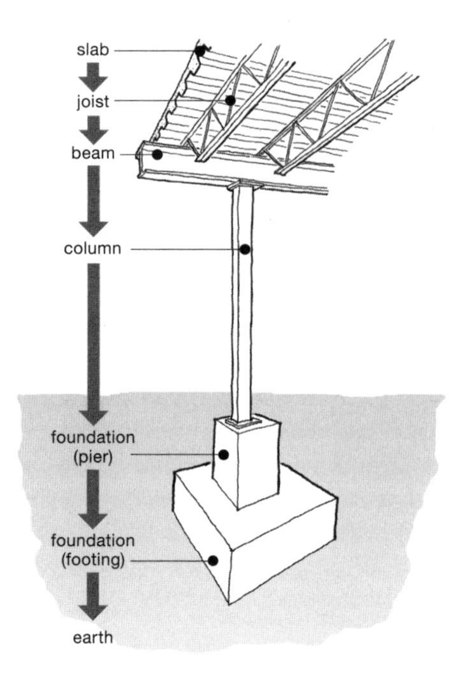
slab
joist
beam
column
foundation
(pier)
foundation
(footing)
earth

Make sure it doesn't work the wrong way.

The downward transfer of structural forces through a building is the **load path.** Loads sometimes follow a path different from the one intended, which can result in structural failure. For example, every beam will sag slightly under normal loading. If a nonstructural partition is built directly under it, the beam may transmit its loads to the partition. The partition may deform, or it might transfer the loads from the beam to the floor below, causing the floor to sag or even fail.

Figuring out how to make a system work is as important as figuring out how to make it not work in undesirable ways.

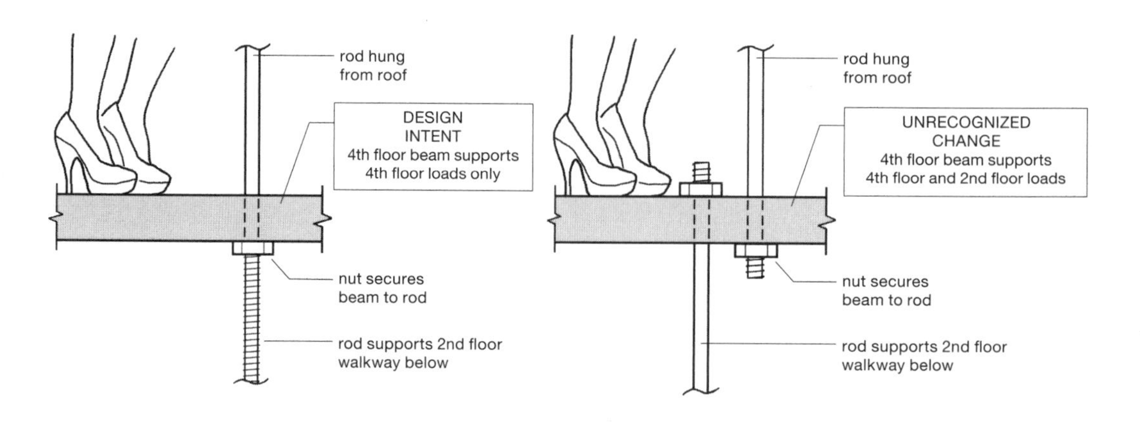
rod hung
from roof
DESIGN
INTENT
4th floor beam supports
4th floor loads only
nut secures
beam to rod
rod supports 2nd floor
walkway below
As designed
rod hung
from roof
UNRECOGNIZED
CHANGE
4th floor beam supports
4th floor and 2nd floor loads
nut secures
beam to rod
rod supports 2nd floor
walkway below
As built

Kansas City Hyatt walkway collapse

On July 17, 1981, two interior atrium walkways collapsed during a dance party at the Hyatt Hotel in Kansas City, Missouri. One hundred fourteen people were killed and more than 200 were injured.

The walkways crossed the multistory atrium at the 2nd and 4th floors, and were suspended from the roof by steel rods. The engineer intended the rods to be continuous, with the upper walkway held in place by nuts and the rods continuing to the walkway below.

During construction, the fabricator recognized the difficulty of installing 4 story–long threaded rods, and of rotating the nuts two stories into place. A proposal for two sets of shorter rods was put forth. One set would hang the 4th floor walkway from the roof, and a second set would hang the 2nd floor walkway from the 4th floor walkway. The engineer approved the design change without performing a structural analysis.

Post-accident analysis revealed that the revision had doubled the load on the 4th floor steel beams. Further, each beam specified by the engineer was not a single member, but two parallel members welded together. Under full loading, the welds failed, causing the upper walkway to "pancake" onto the lower walkway.

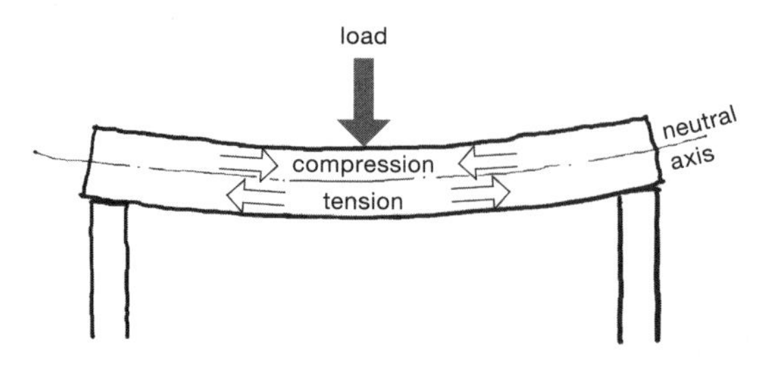
load
compression
tension
neutral
axis

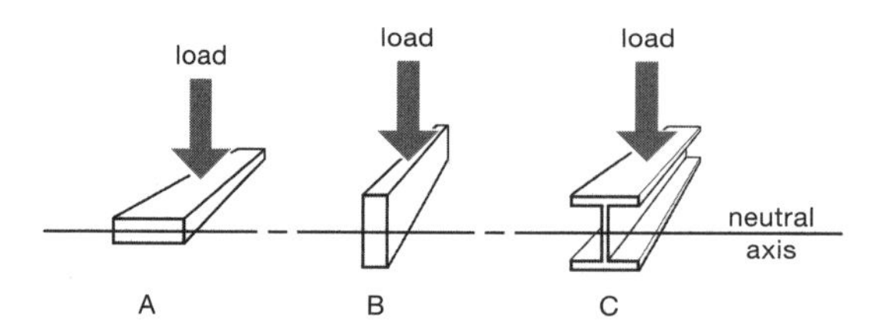
load
load
load
neutral
axis
A
B
C

The best beam shape is an I—or better yet, an I.

When a beam bends under loading, the top shortens (is compressed) and the bottom stretches (is tensioned). Compressive and tensile stresses are respectively greatest at the very top and bottom of the beam, and they decrease toward the middle, where stresses are zero at the **neutral axis.**

A rectangular beam used in a vertical orientation (B) is more effective than the same beam used horizontally (A), because a greater percentage of its material is located away from the neutral axis, where most of the beam's work needs to be done. For the same reason, an I-shaped beam (C) is even more efficient.

Simple span with beam depth d

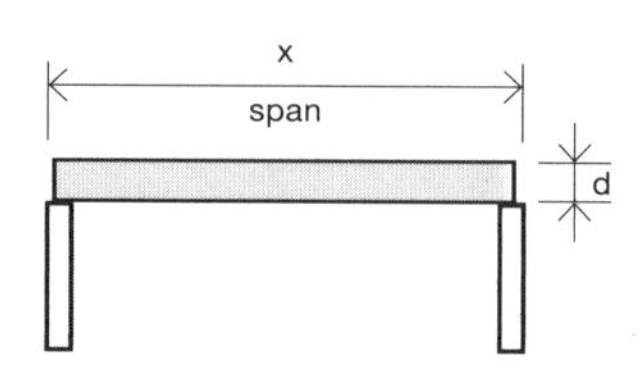

Most beams can be cantilevered a short distance without increasing depth.

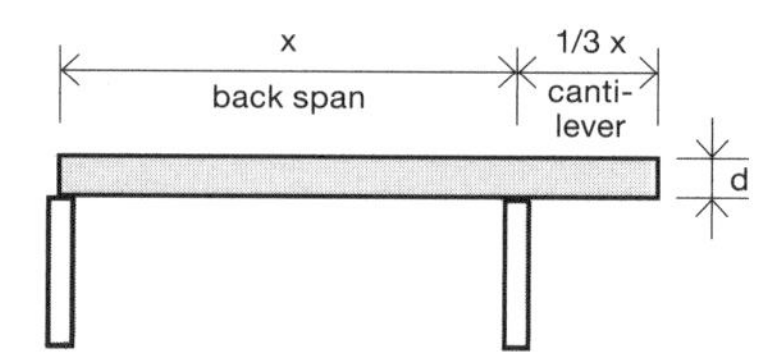

A beam of a given length can be made shallower by moving a column to create a back span and cantilever.

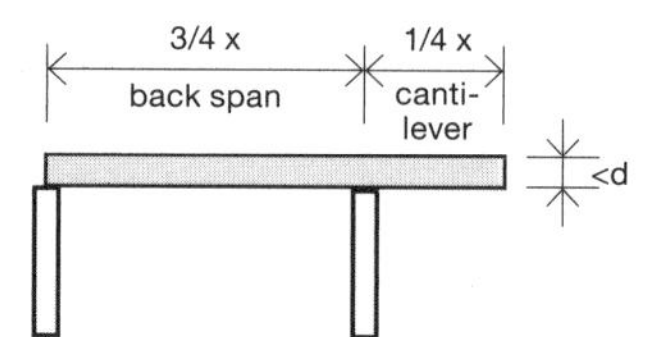

Get even more out of a beam.

A useful measure of a beam's efficiency is the ratio between the distance it spans and the amount of material it uses. Efficiency can be improved in several ways.

Use cantilevers. Extending a beam beyond its support helps spring the back span upward. This often allows the use of a beam with less depth than would otherwise be required. Most beams can be safely cantilevered about 1/3 the length of the back span.

Convert point loads to multiple loads. A beam more easily resists a given load if it is distributed among several locations rather than placed at one point. Moving a load from the center of a span toward one end also helps; loads from high in a building sometimes can be redistributed this way to reduce the size of beams below.

Use a truss. Although a truss must be deeper (taller) than a solid beam to do the same work, it uses a fraction of the material.

Cut holes in it. Under many loading conditions, material can be removed from the web (the vertical center portion) of a steel beam, which lightens it while allowing the passage of ducts, pipes, and wires.

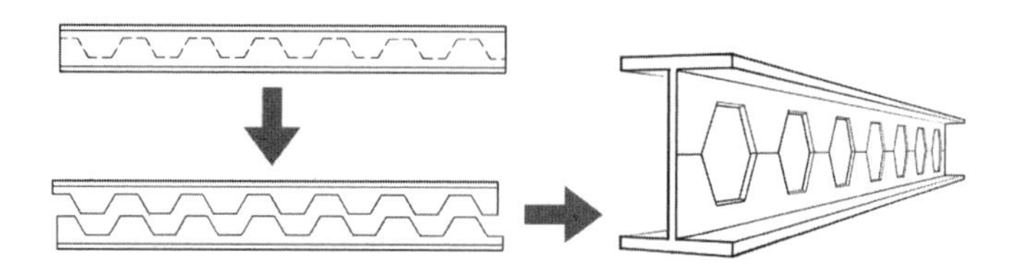

Castellated beam

38

“Inventing is the mixing of brains and materials. The more brains you use, the less materials you need.”

—CHARLES KETTERING

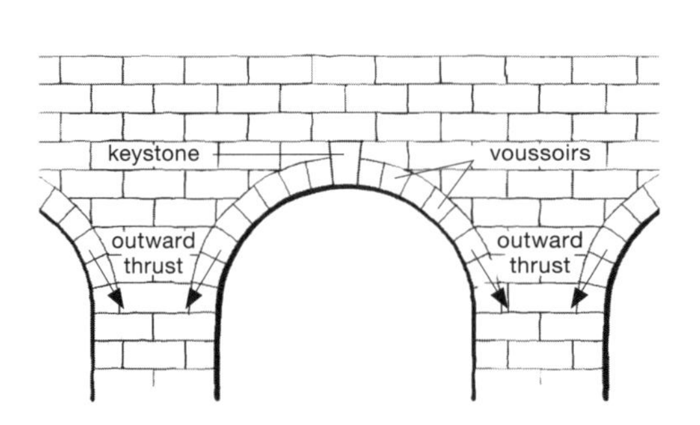
keystone
voussoirs
outward thrust
outward thrust

A masonry arch gets stronger as it does more work.

Gravity is usually the enemy of structural endeavor, as it seeks to pull structures toward the earth. But a masonry arch works *because* of gravity. Gravity pulls each **voussoir** (voo-SWAH) in an arch into contact with the one below it, which transmits the force to the one below it, and so on. The greater the loads on an arch, the more the masonry units cohere, until the compressive strength of the material is exceeded. For this reason, a masonry arch will tend to destabilize when there is a relatively small load on it, or may simply *look* unstable if there is relatively little masonry above it.

At its base, an arch generates an outward thrust in addition to a vertical gravitational force. This must be resisted by a large mass, such as a concrete or earthen embankment for a bridge arch or a "flying" buttress for a large church arch. When arches are placed in series, the outward thrust from each arch neutralizes the thrust from the adjacent one.

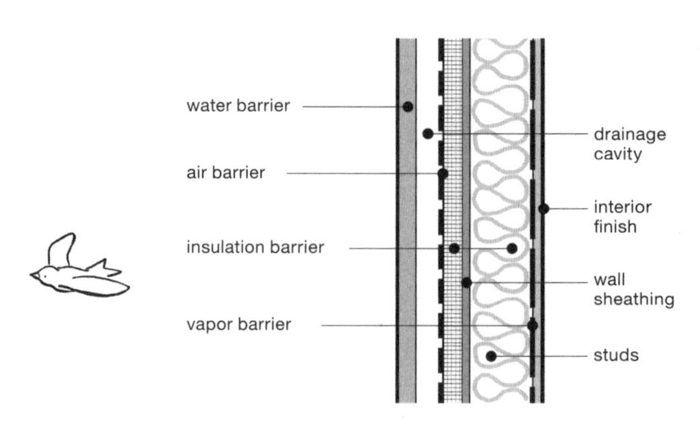

Wood frame wall

The four eras of the wall

Great Mass Era: From early civilization into the late 19th or early 20th century, very thick stone, brick, log, adobe, and concrete walls provided defense against heat, cold, wind, earthquakes, and intruders, as well as primary structural support in buildings.

Curtain Wall Era: Upon the emergence of iron, steel, and concrete structural systems in the late 19th century, the exterior wall shed its load-bearing role and served primarily to enclose interior space.

Insulation Era: The invention of fiberglass insulation in 1938 allowed a 4" wall to offer similar protection from heat and cold as a 2' thick masonry or adobe wall. But as we insulated and sealed buildings more and more, they often developed problems with condensation and air quality.

Specialized Layers Era: The state-of-the-art building envelope has four layers that perform as an integrated system: **water barrier** (a cladding and/or membrane that protects against precipitation); **air barrier** (a membrane that minimizes seepage of exterior air into the building); **insulation barrier** (thermally separates interior and exterior); and **vapor barrier** (prevents movement of moist indoor air into a wall or ceiling cavity). Each layer should be continuous around a building.

The first building code

"If a builder has built a house for a man, and has not made it sound, and the house falls and causes the death of its owner, that builder shall be put to death. If it is the owner's son that is killed, the builder's son shall be put to death. If it is the slave of the owner that is killed, the builder shall give slave for slave to the owner of the house. If it ruins goods, the builder shall make compensation for all that has been ruined, and shall re-erect the house from his own means. If a builder builds a house, even though he has not yet completed it; if then the walls seem toppling, the builder must make the walls solid from his own means."

—THE CODE OF HAMMURABI
by King Hammurabi of Babylonia, 1792–1750 BCE

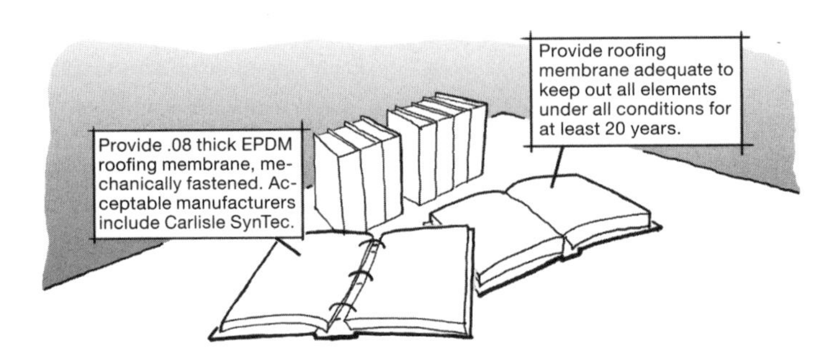

Prescriptive specifications
give detailed characteristics of products or systems to be used, including materials, dimensions, and methods of installation

Performance specifications
identify the desired performance, such as strength, capacity, and stability, without indicating how the contractor is to achieve it

Drawings explain only some things.

Drawings, no matter how detailed, tell only part of the story of what is to be manufactured or built. A separate, comprehensive **specifications** document, also prepared by the engineer, provides details on the **what** (e.g., strength of concrete and steel, acceptable fastener types, wiring gauges), **who** (qualifications of subcontractors, acceptable component manufacturers), **when** (schedules, sequence of tasks, procedures for review of the work), **where** (parts of a construction site that may be used for certain activities), and **how** (handling of materials, methods of applying finishes, etc.).

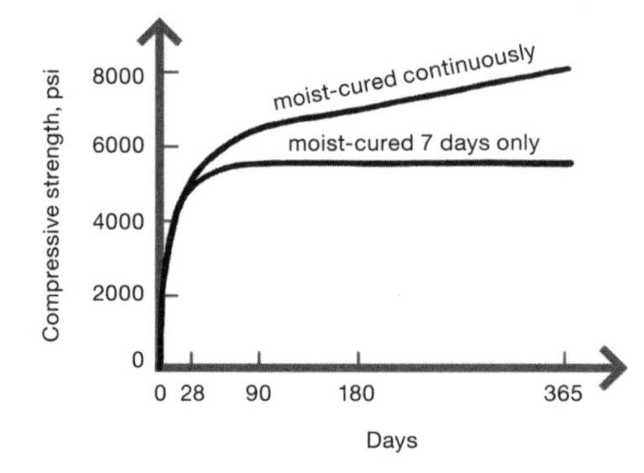

Compressive strength, psi
8000
6000
4000
2000
0
moist-cured continuously
moist-cured 7 days only
0
28
90
180
365
Days

Concrete doesn't dry; it *cures*.

43

Concrete gains strength through a chemical reaction between cement and water. After pouring, concrete is often kept wet (moist-cured) for an extended period to prolong the chemical reaction (thereby strengthening the product), and to keep outer portions of the concrete from drying out long before interior portions (thereby minimizing cracking). Design calculations for concrete construction typically are based on the strength expected after 28 days of curing. However, the maximum strength of a very large pour might not be achieved for decades.

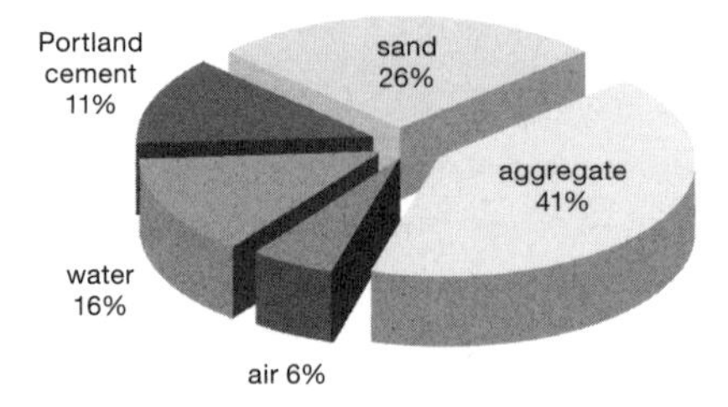

Typical concrete mix design

Concrete and cement are different things.

Cement is a binding and hardening ingredient, and is primarily derived from limestone. It is mixed with sand, aggregate (rocks or pebbles), air, and water to make concrete. Chemicals may also be introduced into a concrete mixture to speed or slow the hardening of the concrete, make it heavier or lighter, or enhance resistance to environmental factors.

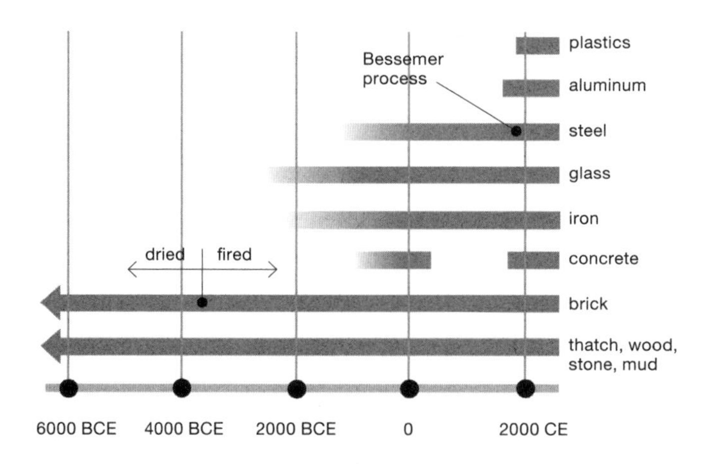
plastics
Bessemer process
aluminum
steel
glass
iron
dried
fired
concrete
brick
thatch, wood, stone, mud
6000 BCE
4000 BCE
2000 BCE
0
2000 CE

Concrete and steel are ancient, not modern, materials.

It was known many centuries ago that a small amount of carbon made iron into a stronger metal, but mass production of steel was not possible until the invention of the Bessemer process in the 1850s. Concrete was used in the Roman Empire, but after the empire's dissolution the technology was lost until the 18th to 19th centuries.

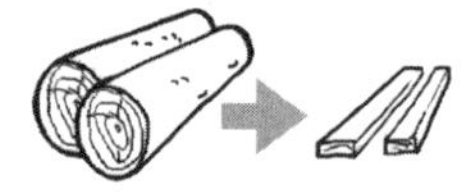

Separating
material is sized and shaped by removing excess

Casting/molding
a molten or liquid material is placed in a mold to solidify

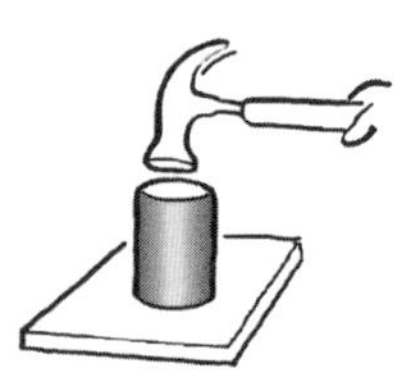

Forming
material is shaped via a die (a shaped metal block)

Conditioning
material properties are changed through heat, pressure, or chemicals

Assembling
individual pieces are combined, e.g., on an assembly line or in a garment factory

Finishing
surfaces are protected or beautified by tempering, coating, decorating, etc.

Secondary processing

3 stages of manufacturing

Material extraction: Raw materials, such as trees, crops, oil, and minerals, are identified and harvested.

Primary processing: Extracted materials are put into standard formats for use by industry. Limestone, sandstone, and shale are baked and crushed into powder to make cement; alumina is extracted from bauxite ore, processed, and cast into aluminum ingots; cotton is cleaned, deseeded, and compressed into bales; grain is ground into flour.

Secondary processing: Primary industrial materials are made into products for use by consumers.

Judgment by inspector		Actual Condition: part is defective	Actual Condition: part is not defective
	part judged defective	correct assessment	**Type 1 error** false positive
	part judged not defective	**Type 2 error** false positive	correct assessment

More inspections and fewer inspections both produce more errors.

Inspection occasionally rejects a good item or fails to identify a defective item. A **false positive** has little consequence other than the cost of replacing the item. But a **false negative** can have great consequence, as the item may fail after being placed in service.

More inspections are not necessarily the answer, however. Statistically, the addition of an infinite number of inspections will cause nearly every item to be found defective for some reason. An optimal level of inspection balances the economics of replacing false positives with the human and moral consequences of failing to detect real errors.

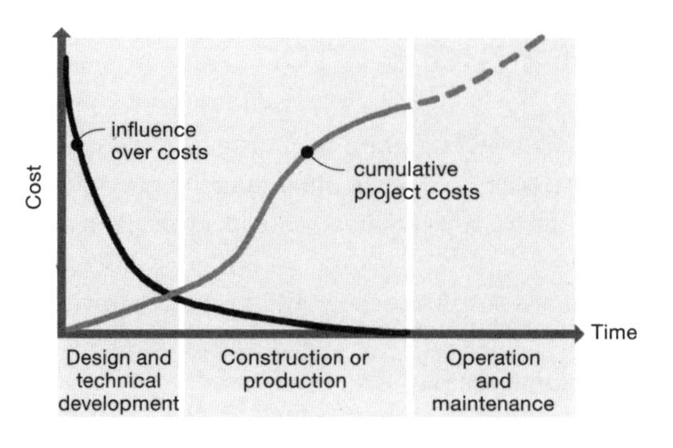
Cost
influence
over costs
cumulative
project costs
Time
Design and
technical
development
Construction or
production
Operation
and
maintenance

Early decisions have the greatest impact.

Decisions and assumptions made just days or weeks into a project—regarding end-user needs, scheduling, the size and shape of a building footprint, and so on—have the most significant impact on design, feasibility, and cost. As decisions are made later and later in the design process, their influence decreases. Minor cost savings sometimes can be realized through **value engineering** in the latter stages of design, but the biggest cost factors are embedded at the outset, in a project's DNA.

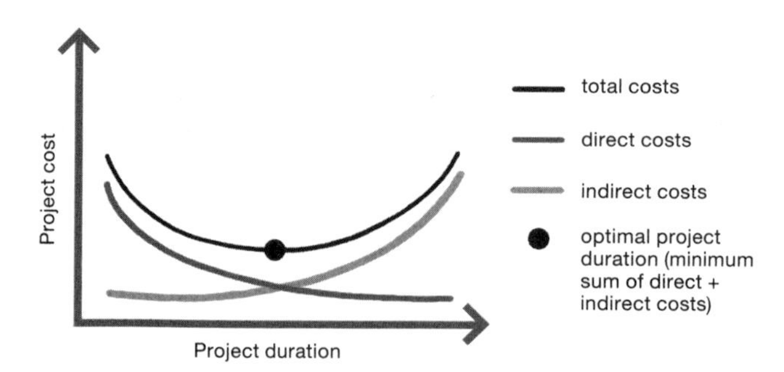

Project cost
Project duration
total costs
direct costs
indirect costs
optimal project duration (minimum sum of direct + indirect costs)

Working faster doesn't save money.

When a production schedule is accelerated, savings are often expected through reduced **indirect costs**—overhead, equipment rental, insurance, supervision, utilities, and so on. Meanwhile, **direct costs** (generally, labor, materials, and equipment purchases and operation) may be expected to stay constant, because the same amount of work must be done regardless of schedule.

In practice, however, faster work produces more confusion, errors, substandard quality, and overtime pay, driving up costs. An extremely long work period also increases total costs, particularly indirect costs. **Optimal project duration** minimizes the combination of indirect and direct costs.

Occasionally, the greater costs of an accelerated production schedule are acceptable. In a highly profitable real estate market, a developer eager to make a building available for lease might employ **fast-tracking,** in which construction begins before the building is fully designed. This drives up costs for the above reasons, and also because many parts of the building, such as the foundation and structural system, must be overbuilt to allow for the worst-case outcome of design decisions not yet made.

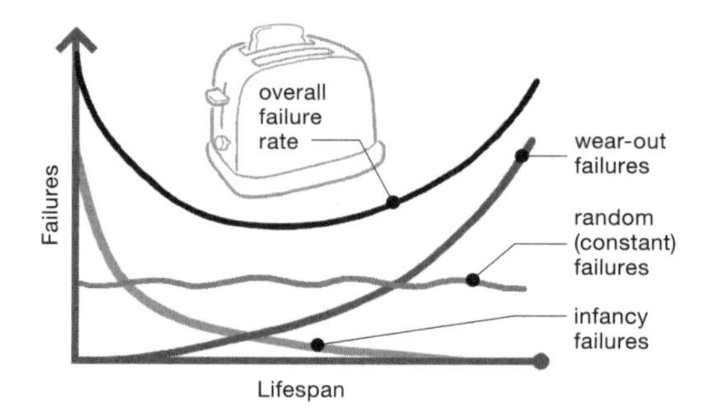

Common "bathtub curve" of reliability for many products

Perfect reliability isn't always desirable.

Reliability is the measure of how long a product or system functions properly. **Target reliability** is expressed as a number between 0 and 1. A target reliability of 1 indicates a goal of perfect reliability; a target reliability of 0 would indicate a goal of all failure. The target reliability of a bridge, spacecraft, pacemaker, or similarly critical system is 1, because failure may result in loss of life. Products that are relatively inexpensive, such as toys or DVD players, are designed with a target reliability of less than 1, because failure is not critical and the cost of achieving perfection would increase cost. Surprisingly, some aircraft parts have a target reliability of less than 1, because of the need to minimize weight. This is mitigated by routine replacements and frequent inspections to identify potential failures.

Failures occur for different reasons over the life of a product or system, with ordinary wear-out failures eventually overtaking start-up (infancy) failures.

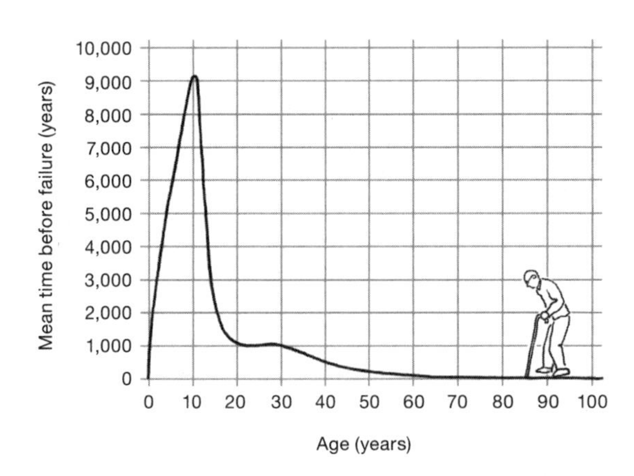
Mean time before failure (years)
10,000
9,000
8,000
7,000
6,000
5,000
4,000
3,000
2,000
1,000
0
0
10
20
30
40
50
60
70
80
90
100
Age (years)

Human MTBF

Human time to failure is 1,000 years.

Mean time before failure is the inverse of the expected failure rate of a device or system. A 25-year-old person has an MTBF of about 1,000 years, because the annual rate of death (failure) for a person that age is 1 in 1,000, or 1/1,000. As we age and near the end of our **service life,** our MTBF decreases. There is no direct correlation between service life and failure rate. A rocket is designed to have an MTBF of several million hours, because failure would be catastrophic. However, its intended service life is only a few minutes, such as during the launch of a spaceship.

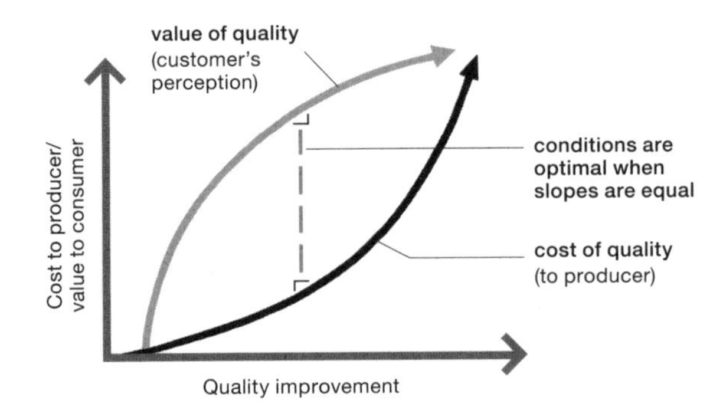

The quality-cost curve

Few customers will pay for a perfectly engineered product.

Customers notice and are willing to pay for improvements to low-quality products more than high-quality products. A 10% improvement to a low-quality product will lend more than a 10% increase in the **value of quality**—the user's perception of its quality. But as subsequent improvements are made, they add value at a decreasing rate. If a 10% quality improvement costs $10, a 20% improvement will cost more than $20. Eventually, the **cost of improving quality** increases at a faster rate than the improvement will be perceived.

The **optimal quality-cost state** theoretically occurs when the slopes of the value and cost curves are equal. At this point, the rate of improving a product equals the rate at which costs to the producer will increase. Beyond this point, the producer's cost for providing one more unit of quality will exceed the value the customer will perceive.

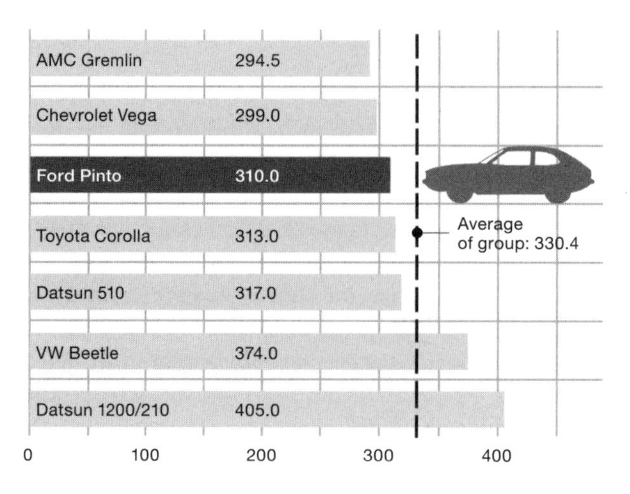

Average annual occupant fatalities per million vehicles, 1975–76

The Ford Pinto wasn't unsafe.

Following the influx of small foreign cars into the U.S. in the 1960s, Ford developed the Pinto on an accelerated schedule. Shortly after its release, the vehicle was assailed for catching fire in rear-end crashes. Over 500 deaths were said to have resulted from design flaws such as bolts protruding from the rear differential near the gas tank.

In a wrongful-death lawsuit, an internal Ford document surfaced that stated that unimplemented improvements to the gas tanks would have cost only $11 per vehicle. Ford, using a human life value of $200,000, determined it would cost far less to pay for injuries and deaths than to improve 12.5 million vehicles. The legal standard of the day was expected to excuse Ford from liability, because the courts to that point had not considered a defendant negligent if the cost of an improvement exceeded its benefit. But the jury found Ford liable and ordered it to pay $3M compensation and $125M in punitive damages (later reduced to $3.5M).

A later study revealed that the unimplemented $11 improvement was never meant to address gas tank failures in rear-end collisions. Nor did Ford place a $200,000 value on human life; this value was created by the National Highway Traffic Safety Administration. Statistics indicate the Pinto's overall safety record was average for its day, with its registration rate matching its rate of involvement in vehicle fatalities.

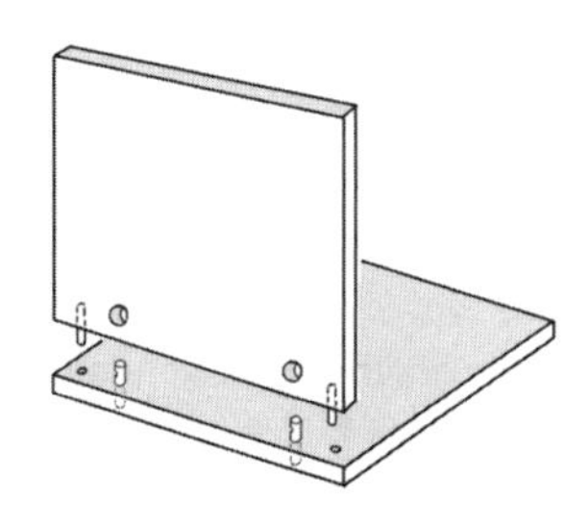

Home furniture assembly

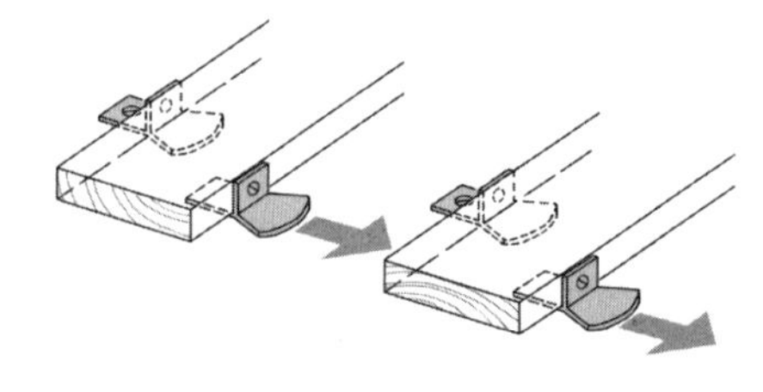

Hidden fasteners for porch boards

Be careful when asking a part to do more than one thing.

It may seem desirable to minimize effort, material, and time by having one feature or part serve multiple purposes. The viability of this approach depends on the level of skill and care that can be expected during use or application. The greater the sophistication of the end user and the more controlled the user's environment, the more a designer may rely on multi-functionality. Where an error would be catastrophic, it is usually better to have each part serve only one purpose.

IKEA furniture frequently employs one set of hardware to align parts, and a separate set to fasten them together. Each set serves one purpose, minimizing opportunities for error by the home assembler.

Hidden porch board fasteners are clips that serve two purposes, but only one purpose at a time. On one side of a board, they are fastened to the structure below. On the other side, they are tucked under the previously installed board. If two clip types were provided, one could easily install the wrong clip.

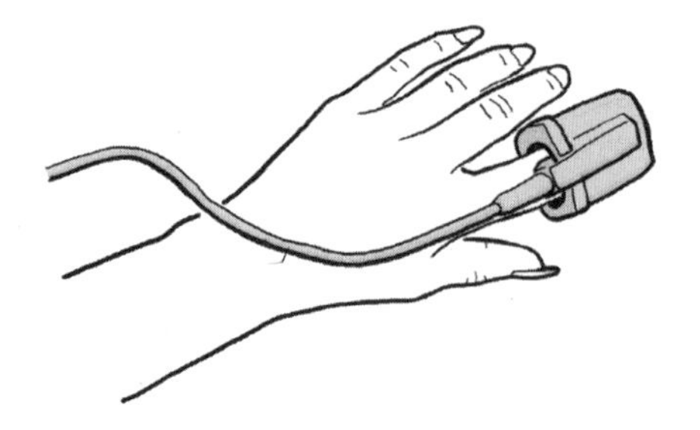

Design a part to fail.

Electrical systems are protected by fuses or circuit breakers that trip before a power surge can ruin expensive components or damage hard-to-access wires.

In **steel buildings,** connections between structural members may be designed to deform during earthquakes, to prevent catastrophic failure of the larger system. Repair of the connections can be done at a fraction of the cost of replacing the entire building.

Biomedical devices are often connected loosely to protect a patient. A pulse oximeter, used to detect blood oxygenation, is connected to a patient's finger with deliberate weakness, preventing injury should someone trip over the cable.

The clips that hold **lobster traps** together are designed to corrode after one fishing season. When traps are lost or abandoned, the clips will fail before the wire-gridded sides, leaving flat pieces that are much less hazardous to boats than a pile of submerged boxes.

up to 1mA	tingle sensation
1-2 mA	uncomfortable sensation
5 mA	maximum harmless current
15 mA	maximum “let-go” current
10-20 mA	“can’t let go” current

(1000 mA = 1 Amp)

Electric shock values

Keep one hand in your pocket.

If one of your hands is touching any object while the other hand touches a piece of electrical equipment that releases an errant charge, the charge may seek a path to the ground, from one hand to the other, through your heart. Keeping one hand in your pocket won't keep you from getting shocked, but it will encourage a charge to follow a less dangerous path—through your hand, up your arm, and down the nearest leg to the ground.

Keep one leg still.

To level a surveying tripod, set it in the desired location at approximate level. Repeatedly adjust two of the legs and ignore the third leg until the bull's-eye indicates dead level.

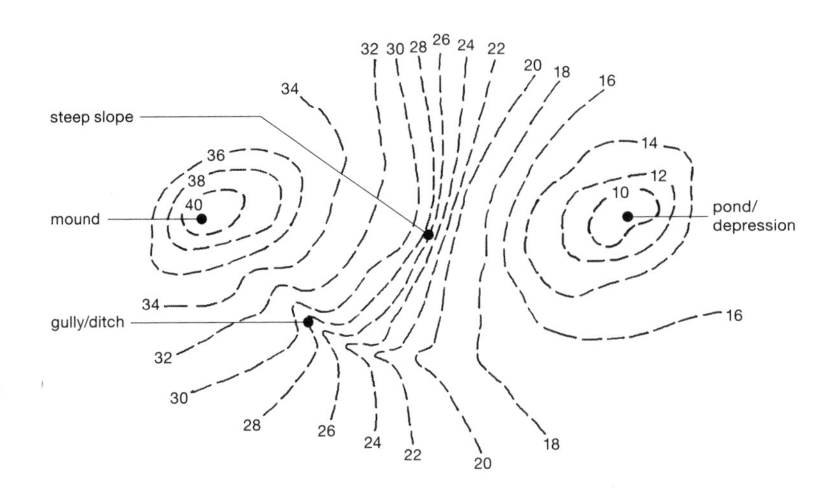

steep slope
mound
gully/ditch
pond/
depression
40
38
36
34
34
32
32
30
30
28
28
26
26
24
24
22
22
20
20
18
18
16
16
14
12
10

How to read a topographic plan

A topographic plan depicts a landscape through a series of contour lines. Each line indicates a constant **elevation**—a measured height above sea level or other reference point. Several keys help in reading a topographic plan:

- The direction of slope is perpendicular to the contours. A meandering drop of rainwater moves perpendicular to the contour lines, from higher to lower elevation.
- Where contour lines are closely spaced, the terrain is steep; where farther apart, flatter.
- If a nearby lake were to flood to a given elevation, the outline of the lake would match a contour line.
- If discerning a ridge (or crown) from a gully (ditch) is difficult, place yourself on the plan at the edge of the ridge or gully, and imagine yourself walking straight across it. At each step, verify the elevation on the plan to determine if you are going up or down.

fill
cut
Existing
Proposed

Balance cut and fill.

Design sitework to equalize the amount of earth to be removed (**cut**) and the amount to be added (**fill**). This simplifies earthmoving and grading, and minimizes the expense of moving soil to and from a construction site.

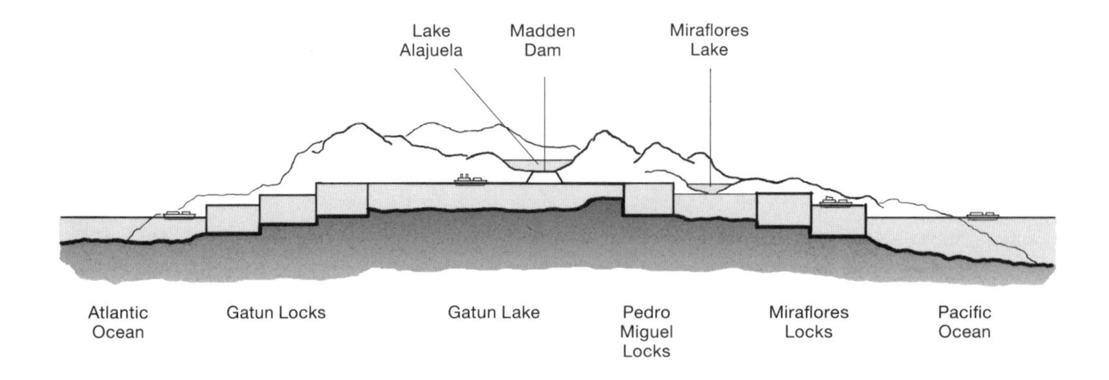

The Panama Canal, schematic section looking northeast

Work with the natural order.

Every ship passing through the Panama Canal must be raised 85' over the natural terrain, and lowered 85' to meet the ocean on the other side. This feat is accomplished without any pumps. Gravity moves millions of gallons of water from mountain lakes down to the lock chambers. As long as precipitation refills the lakes, the locks continue to function.

Turbulent flow

particle paths are irregular;
tends to occur in larger pathways
and at high flow rates

Laminar flow

particles move in straight lines;
tends to occur in small pathways
and at low flow velocities

Two types of flow in fluids

Air is a fluid.

A fluid is any amorphous substance that yields easily to external pressure and assumes the shape of its container. This includes all gases and liquids.

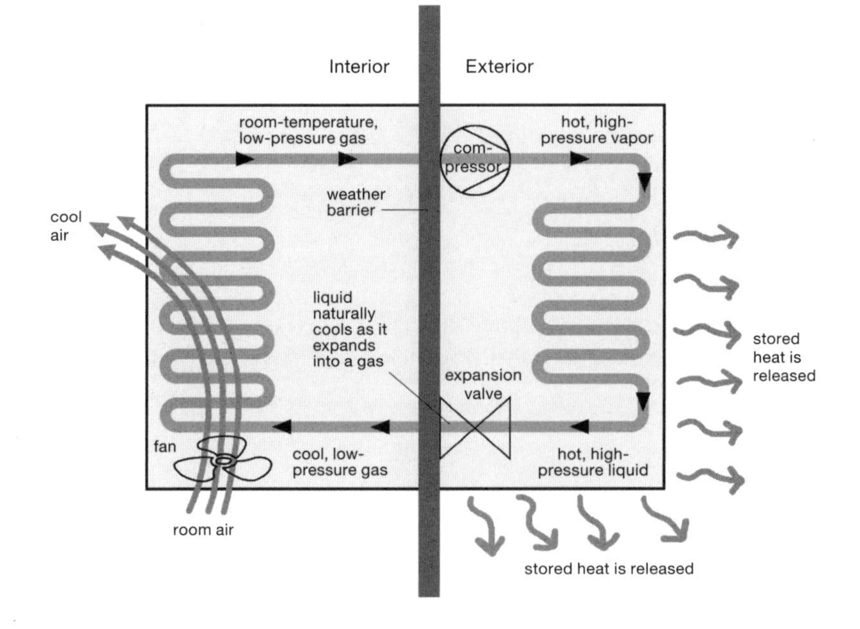

An air conditioner

Heat cannot be destroyed, and cold cannot be created.

An air conditioner doesn't create cold; it moves heat from a building interior to the exterior. It does this by exploiting a natural principle: substances absorb heat when moving from a liquid phase to a gas phase, and release heat when moving from gas to liquid. Central air-conditioning systems, window air conditioners, and food refrigerators work the same way but at different scales. A heat pump is, conceptually, an air conditioner working in the opposite direction, removing heat from outdoor air and moving it into the building interior.

Conduction
heat transfer through direct material contact

Convection
heat transfer through movement of a fluid (air or liquid)

Radiation
energy transfer through space

A radiator doesn't just radiate.

Heat is the movement of molecules within a material. The greater the rate of movement, the greater the heat. Heat is transferred by:

Conduction: When two objects of different temperatures are in contact, or when two areas of one object are at different temperatures, the more active molecules in warmer areas "nudge" the molecules in the colder areas until all molecules are moving at the same speed.

Convection: In gases and liquids, the molecules of the warmer material naturally spread out and move throughout the colder material, losing their heat energy to it. In this way a radiator convects, not just radiates, heat to a room.

Radiation is the movement of electromagnetic waves through space, such as light waves from the sun. The waves provide energy to the molecules that they contact, causing them to become more active and converting electromagnetic energy into heat energy. All matter emits thermal radiation; most can't be felt.

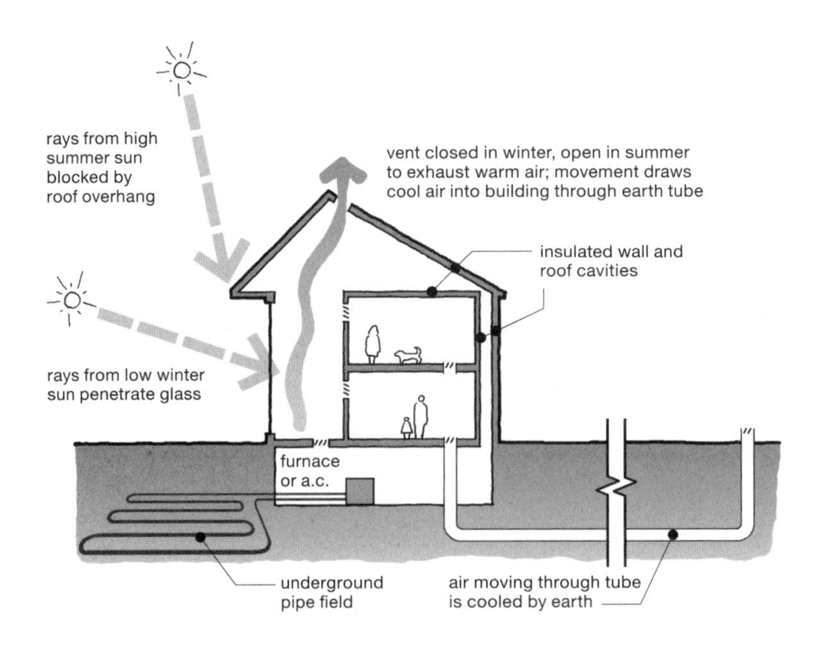

A double envelope house with earth-derived heating and cooling

The most reliable source of heating and cooling is the earth.

Several feet below the surface of the earth, the temperature is more moderate than the air temperature—cooler in the summer and warmer in the winter. In the same way an air conditioner transfers heat from a building interior to the exterior air, or a heat pump transfers heat from exterior to interior, heat can be moved between the earth and an interior space. Hot or cold water circulated through underground pipes will always return to the building at or near the earth's temperature.

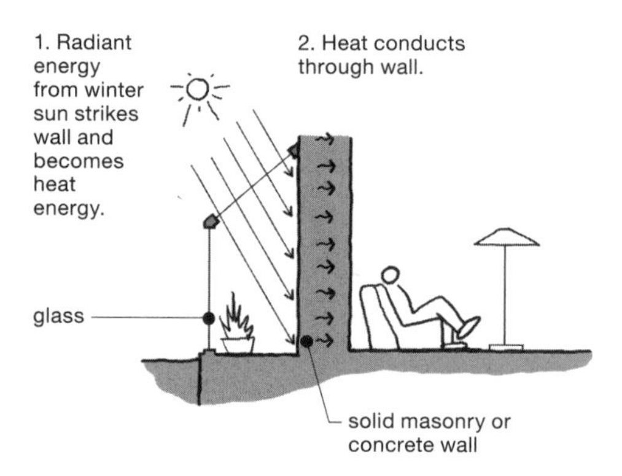

A thermal storage wall

Available solar energy is 50,000 times our energy need.

At least 100 watts of energy strike each square foot of the earth's surface in a fully sunlit hour. Most areas of the U.S. receive the equivalent of 4 hours' full sunlight per day, translating into about 1.5 trillion TWh (terawatt hours) of energy per year—many times the 28,000 TWh used in the U.S.

However, solar cells today can capture only about 20% of the sun's energy that strikes them and are subject to a theoretical maximum of about 33%. And as the percentage of land that feasibly can be covered with solar collectors is small, it is difficult to meet all our energy demands through solar power. At present levels, the U.S. would need a continuous field of solar collectors covering the entire land area of Indiana. If the world used energy at the per capita rate of the U.S., a field the size of Venezuela would be required.

offset impervious
(hard) surfaces

reduce erosion by
minimizing rapid runoff

reduce load on
storm sewer system

support wildlife

reduce aesthetic
impact of development

Benefits of retention ponds

The environmental engineering paradigm shift

Responsibility is to stakeholders, not shareholders. Every creature and every part of the natural environment is a stakeholder in every project.

66

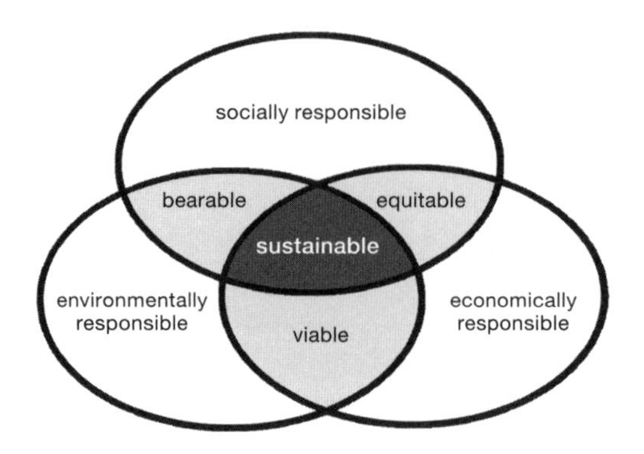

Adapted from John Elkington, *Cannibals with Forks*

Ten Commandments for Environmental Engineers

1 Identify and promote the sustainable use of resources while maintaining a balance of social, economic, and environmental responsibility.
2 Provide safe, palatable drinking water.
3 Collect, treat, and discharge wastewater responsibly.
4 Collect, treat, and discharge human refuse responsibly to prevent disease, fire, and aesthetic insult.
5 Collect, treat, and discharge hazardous materials responsibly to prevent endangerment of human, plant, and animal life.
6 Control and treat air pollutants to reduce acid rain, ozone pollution, and global warming.
7 Design bioreactors to produce biofuels and electricity from organic waste.
8 Design physical, chemical, and biological processes to clean up contaminated sites.
9 Support and enforce the legal regulation of pollutants.
10 Study the transport and fate of chemical pollutants.

With regards to Venkataramana Gadhamshetty

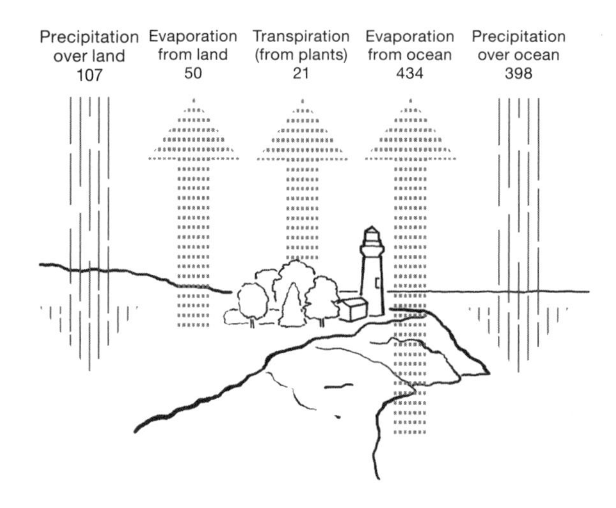

Approximate annual global water movement in 1,000 km^3

Water is constant.

Water moves continuously below, on, and above the surface of the earth. Individual water molecules may come and go quickly or very slowly, but the overall amount of water remains fairly constant.

Reservoir	Average residence
Atmosphere	9 days
Soil	1 to 2 months
Seasonal snow cover	2 to 6 months
Rivers	2 to 6 months
Glaciers	20 to 100 years
Lakes	50 to 100 years
Groundwater, shallow	100 to 200 years
Groundwater, deep	10,000 years
Polar ice sheets	10,000 to 1,000,000 years

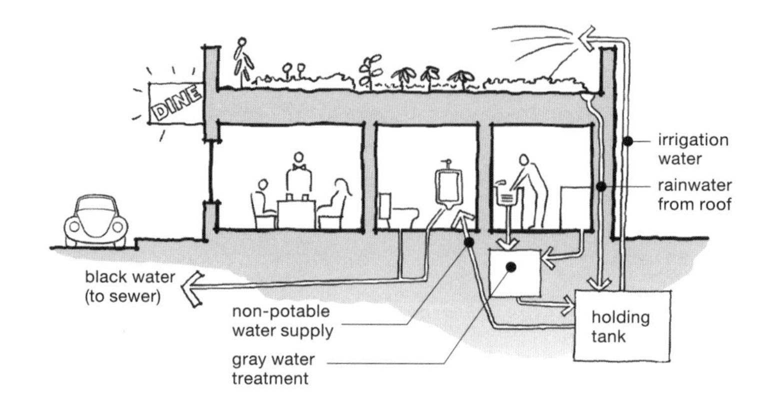

Water recycling

Water recycling

Black water has come into contact with fecal matter. It is not suitable for reuse without extensive treatment, typically via a municipal purification system.

Gray water is wastewater from bathing, cooking, washing, and mild cleaning, and has not come into contact with fecal matter. It is not safe to drink, but with some treatment is suitable for use in toilets and sometimes for the irrigation of plants.

White water is potable (consumable) water from a natural source such as a spring, or that has been treated by a municipal or similar purification system.

Rainwater from outdoor surfaces such as roofs may contain some contaminants from birds and chemicals, but generally may be recycled for use in toilets, car washes, evaporative cooling systems, plant irrigation, and sometimes for consumption by livestock.

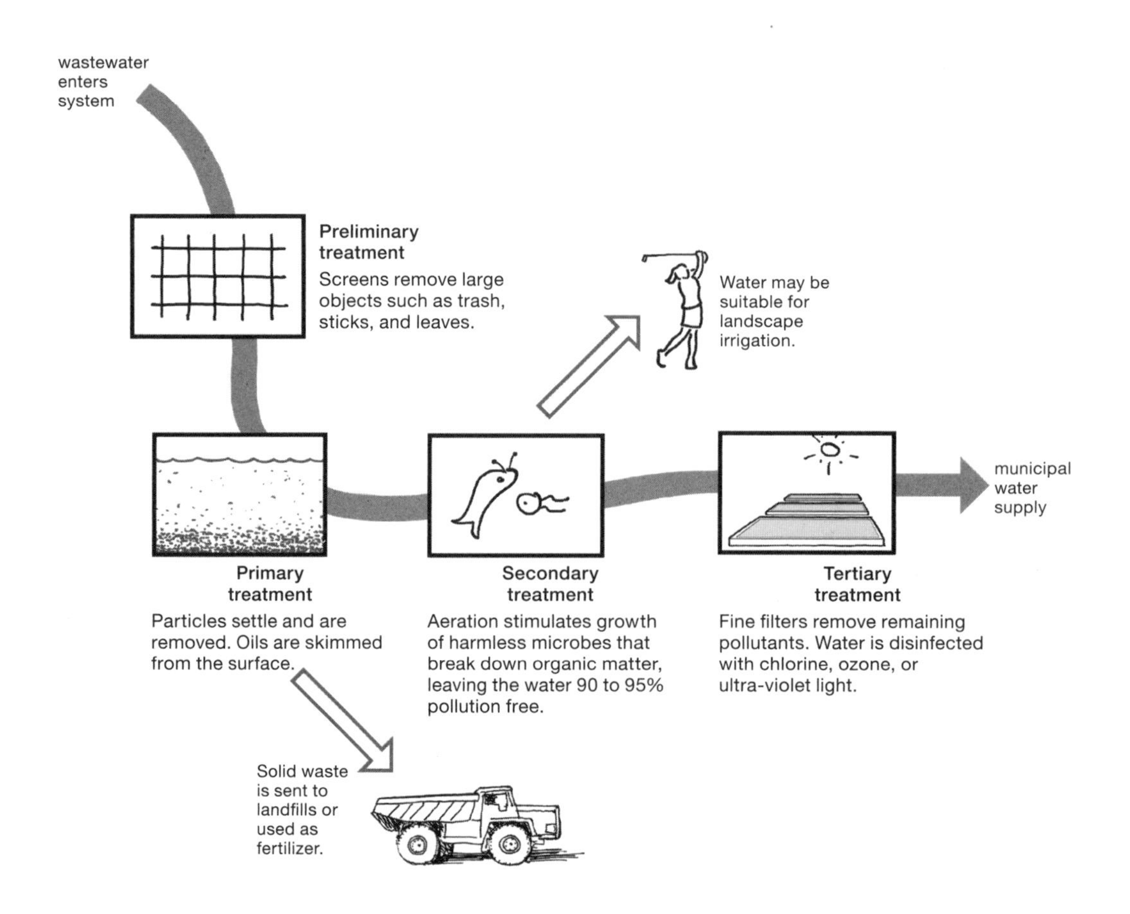
wastewater enters system
Preliminary treatment
Screens remove large objects such as trash, sticks, and leaves.
Water may be suitable for landscape irrigation.
municipal water supply
Primary treatment
Particles settle and are removed. Oils are skimmed from the surface.
Secondary treatment
Aeration stimulates growth of harmless microbes that break down organic matter, leaving the water 90 to 95% pollution free.
Tertiary treatment
Fine filters remove remaining pollutants. Water is disinfected with chlorine, ozone, or ultra-violet light.
Solid waste is sent to landfills or used as fertilizer.

Wastewater treatment imitates nature.

The complex systems used in wastewater treatment plants are accelerated simulations of nature's own purification processes:

- **Settlement basins** approximate a lake.
- **Filtration** approximates settlement to the groundwater table.
- **Aeration** approximates a stream.
- **UV treatment** approximates sunlight.

70

Our community needs
to build a bridge.

The design must
respect our
heritage as a
fishing harbor.

It should have minimal
environmental impact.

It must be in
harmony with
the setting.

TOWN HALL

The problem?

The solution

Don’t presume the solution.

At the point a designer is invited into the design process, many assumptions have often been made about the nature of a problem, its causes, and the desired solutions. The wise designer begins by moving backward—investigating what caused the problem, what caused the causes, and what caused *those* causes. This reveals possibilities that might be very different from what the end user anticipated, but that meet the true need most effectively.

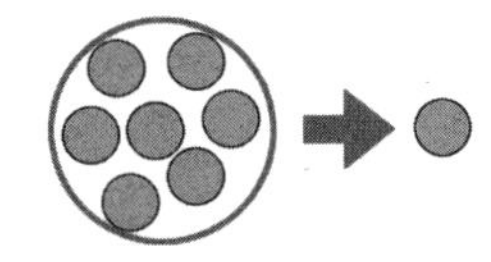

Proper deduction

specific conclusion derived logically from more general truths

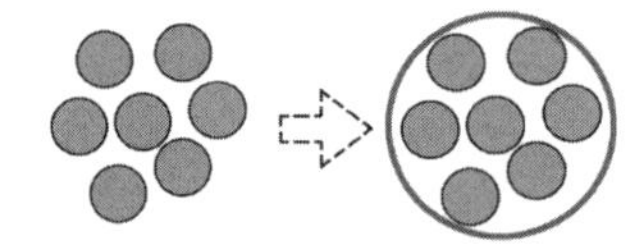

Proper induction

multiple examples suggest, but don't guarantee, a larger truth

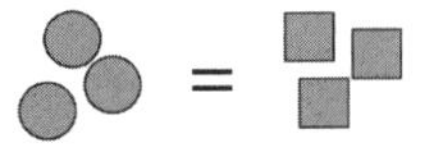

Improper induction

broad generalization or declaration of similarity based on limited data

Think systematically.

Don't congratulate yourself prematurely on an effective piece of analysis. Apply your thinking consistently and thoroughly to all other aspects of the problem at all possible scales, from concept to detail and back again.

Eames molded plywood chair

"We looked at the program and divided it into the essential elements, which turned out to be thirty-odd. And we proceeded methodically to make one hundred studies of each element. At the end of the hundred studies we tried to get the solution for that element that suited the thing best. . . . Then we proceeded to break down all logical combinations of these elements, trying to not erode the quality we had gained in the best of the hundred single elements; and then we took those elements and began to search for the logical combinations of combinations, and several of such stages. . . . And went right on down the procedure. And at the end of that time . . . we really wept, it looked so idiotically simple we thought we'd sort of blown the whole bit. And [we] won the competition."

—**CHARLES EAMES,** furniture designer,
from Ralph Caplan, *By Design*

73

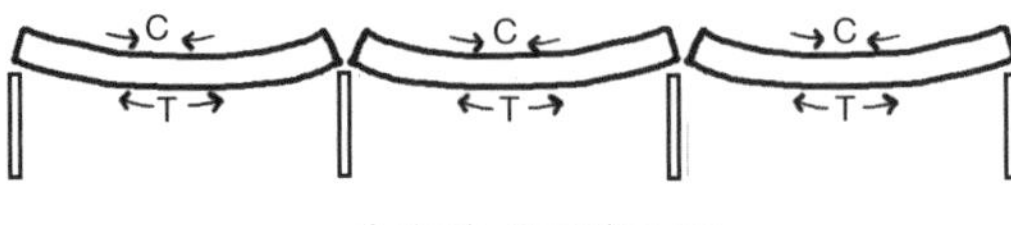

3 single-span beams

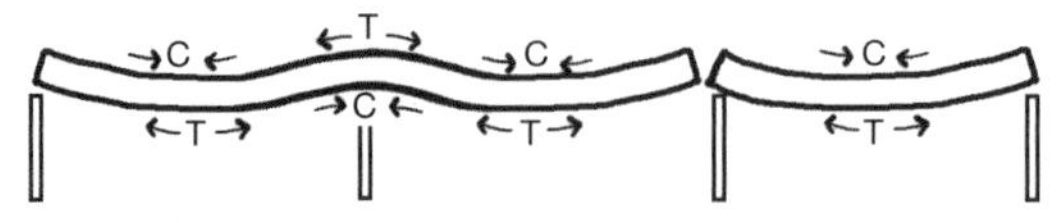

2-span beam + single-span beam

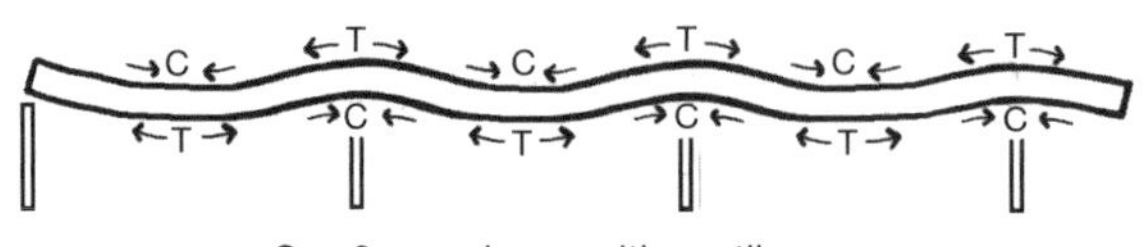

One 3-span beam with cantilever

Tension and compression behavior of beams in three different structural arrangements

Think *systemically.*

A system must be analyzed as a whole, but analysis of the whole is not the summation of the analyses of its parts. The behavior of a part depends upon its relationship to the system in which it resides. And the behavior of the system depends on the many relationships within it, and on the system's relationship to other systems.

Thinking systematically means employing a given thinking method consistently and thoroughly. Thinking *systemically* means thinking about systems and connections—the web of relationships within a system, the relationship of the system to other systems, and the larger system that contains all the systems.

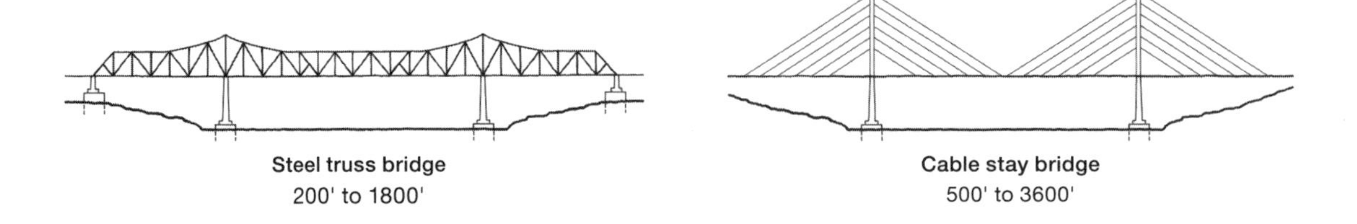

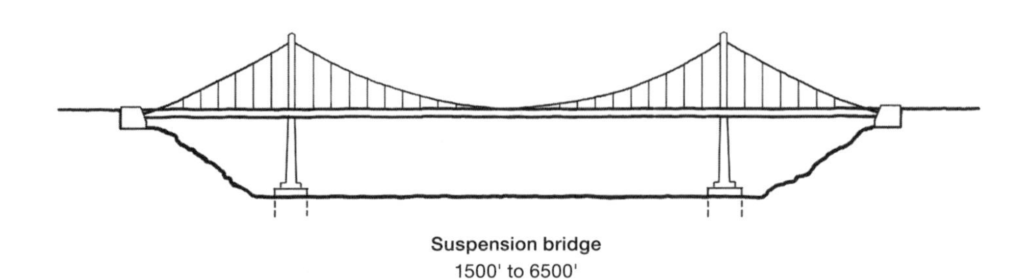

Approximate optimal main spans

A successful system won't necessarily work at a different scale.

An imaginary team of engineers sought to build a "super-horse" that would be twice the size of a normal horse. When they created it, they discovered it to be a troubled, inefficient beast. In addition to being 2 times the height of a normal horse, it was twice as wide and twice as long, resulting in an overall mass 8 times greater than normal. But the cross-sectional area of its veins and arteries was only 4 times that of a normal horse, calling for its heart to work twice as hard. The surface area of its feet was 4 times that of a normal horse, but each foot had to support twice the weight per unit of surface area compared to a normal horse. Ultimately, the sickly animal had to be put down.

Derived from "The Possibility of Life in Other Worlds" by Sir Robert Ball

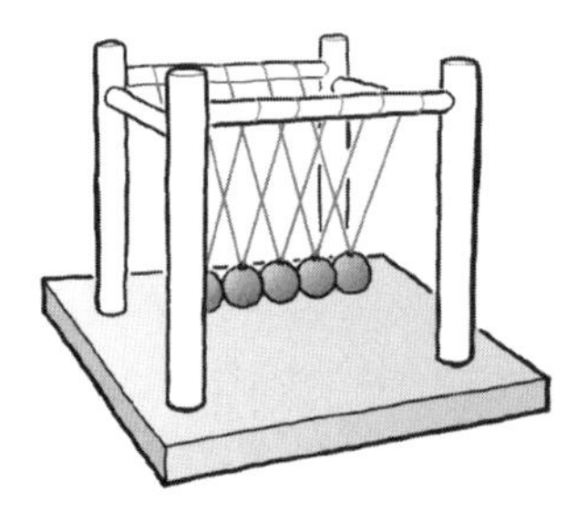

Deterministic system

outcomes can be predicted through known laws and relationships

Stochastic/probabilistic system

outcomes depend on chance or unknown relationships

The behavior of simple systems and complex systems can be predicted. In-between systems: not so much.

The behavior of a lone ball, set in motion by a known force on a billiard table, can be measured or predicted rather accurately. The behaviors of two balls will be more difficult to measure or predict, but the task is somewhat manageable. As the number of objects in the system is further increased—to five, ten, a hundred balls—tracking and/or predicting the behaviors of the individual balls becomes increasingly difficult and eventually impossible.

But at a later point, predictability reenters the model, albeit in a different form. It will remain difficult or impossible to predict the individual behaviors of a million balls on an enormous billiard table, but we will be able to predict many average behaviors, such as how often one ball is likely to collide with other balls, how many balls will strike a side rail in a second, or the average distance a ball will move before striking another.

With regards to Jane Jacobs,
The Death and Life of Great American Cities

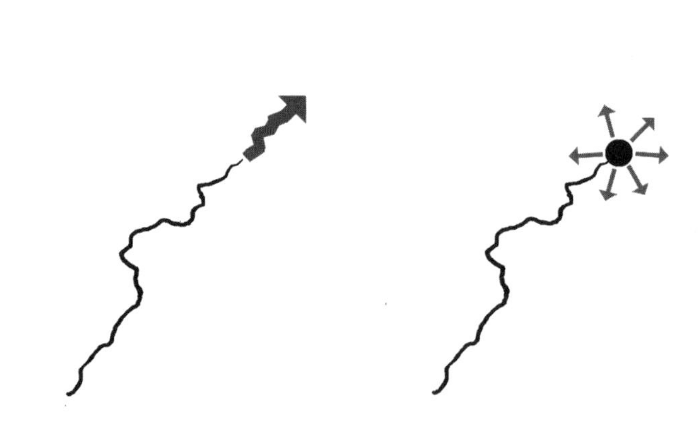

Stop a crack by rounding it off.

Crack propagation in a material increases with the sharpness of the tip of the crack. Drilling a hole at the tip makes a crack less sharp and distributes stresses over a larger area and in more directions, discouraging the crack from lengthening.

Rounding of corners in building products, machine parts, furniture, and even the windows of ships and airplanes provides similar benefit. A rounded window corner spreads stress in multiple directions, while a sharply squared corner directs stress through one point in the system—a crucial consideration in the design of a "thin shell" structure.

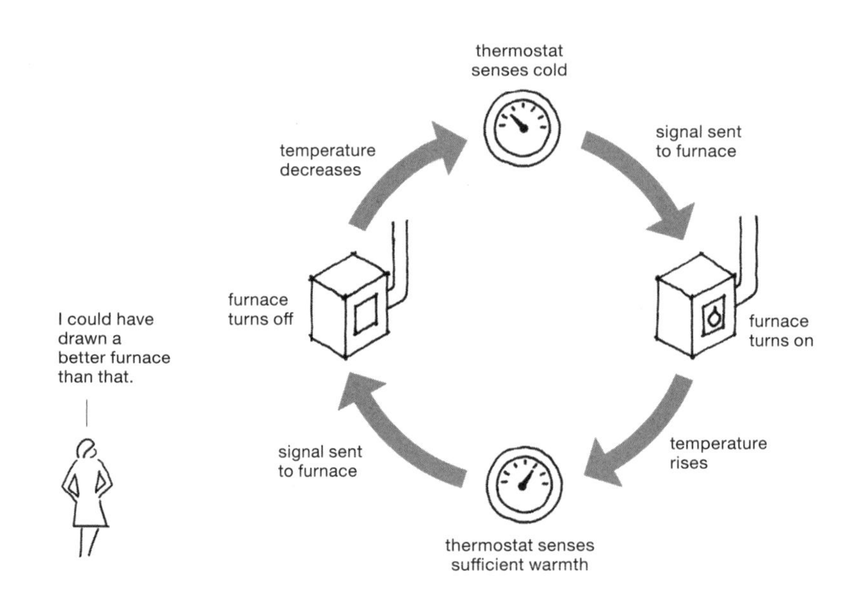

Negative feedback

Seek negative feedback.

In a **negative feedback** loop, a system responds in the opposite direction of a stimulus, providing overall stability or equilibrium. For example, population growth of a species results in overconsumption of the food supply; this leads to a decrease in population, which leads to an increase in available food; this promotes a population increase, and so on until a theoretical equilibrium is attained.

In a **positive feedback** loop, the system responds in the same direction as the stimulus. This decreases equilibrium further and further. For example, a non-native species invades the food supply of a native species; the native species retreats to outlying territory; the non-native species is able to further expand its geographic range, leading to further retreat of the native species.

Most engineered systems rely on negative feedback. In some instances, such as when momentum is desired, a positive feedback loop may be sought.

the problem

|

the cause of the problem

|

the cause of the cause of the problem

|

the cause of the cause of the cause of the problem

|

the cause of the cause of the cause of the cause of the problem

|

the cause of the cause of the cause of the cause of the cause of the problem

Enlarge the problem space.

Almost every problem is larger than it initially appears. Anticipate this by enlarging the problem at the outset—not to make more work, but because the scope of the problem almost certainly will grow on its own. It's easier to reduce the problem space later in the process than to enlarge it after you start down a path toward an inadequate solution.

79

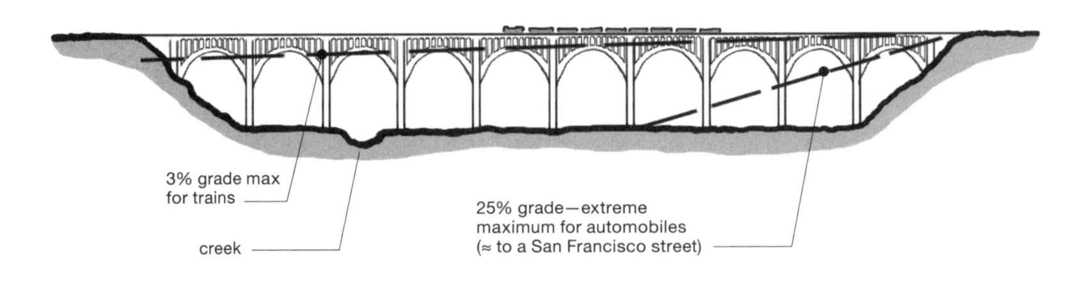

Tunkhannock Viaduct

The Tunkhannock Viaduct

The Delaware, Lackawanna & Western Railroad wished to replace a complex route between Scranton, Pennsylvania, and Binghamton, New York, with a straighter, flatter rail line. Key to its proposed Clarks Summit–Hallstead Cutoff was the crossing of the 75' wide Tunkhannock Creek in Nicholson, Pennsylvania. As the creek was located in a valley too steep for trains to descend and climb, an otherwise modest crossing required a 2,375' long, 240' wide bridge.

Construction began in 1912 and was completed in 1915. Thirteen piers were sunk to bedrock, the deepest of them 12 stories below the surface. The project used 1,140 tons of steel and 167,000 cu. yd. of concrete—enough to cover a football field with a 7 story–high solid mass. The viaduct was, and remained for at least 50 years, the world's largest concrete structure. It remains in daily use.

80

Chemists

investigate chemical interactions and effects; create new solvents, polymers, and pharmaceuticals

Chemical engineers

translate the discoveries of the chemical laboratory into large-scale industrial production

Almost everything is a chemical, and almost every chemical is dangerous.

Chemicals have a limited range in which they are useful to engineers. Below some thresholds of concentration, chemicals cannot be used productively in industry, and beyond a high limit of concentration they are toxic and unmanageable. Even water is a chemical, and it is dangerous in large amounts. Ingestion of too much can alter the body's chemical balance, deplete electrolytes, compromise organ functions, and cause death.

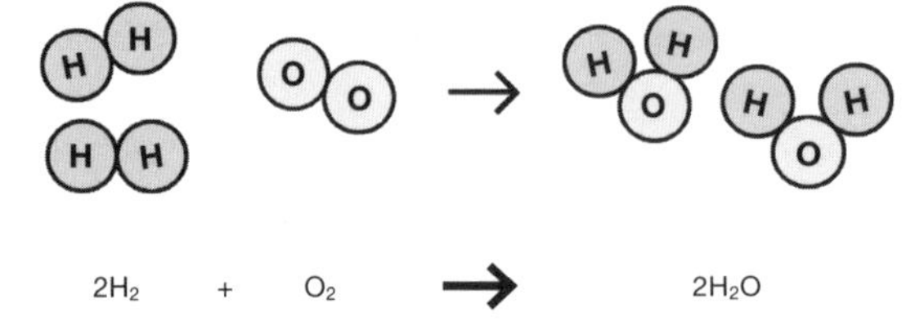

Balancing of hydrogen and oxygen atoms to make water

A chemical equation isn't exactly an equation.

A chemical equation does not represent an equality in a conventional mathematical sense, but indicates the direction and result of a chemical reaction. When reactants are placed together, they interact to form a new compound, or product:

$$\text{reactant} + \text{reactant} \longrightarrow \text{product}$$

In the equation, "+" indicates "reacts with," and "⟶" indicates "produces." However, an equality is present in that atoms are not created or destroyed. The total number of each type of atom before and after a reaction is constant, even though rearranged into new molecules.

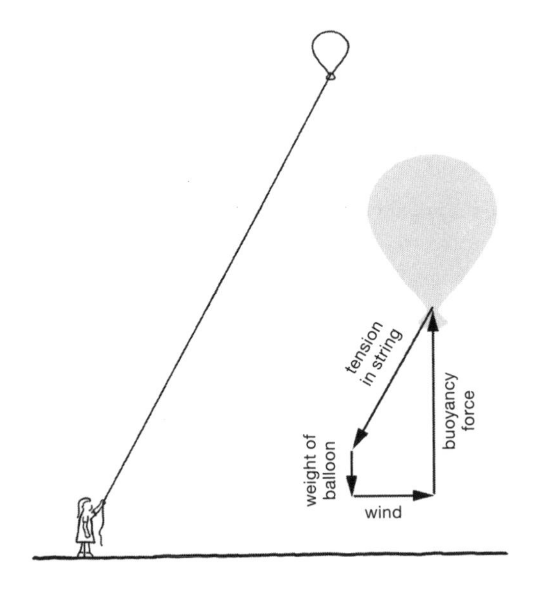

When an object is in equilibrium, the sum of all forces acting on it is zero. The vectors form a closed polygon.

Equilibrium is a dynamic, not static, state.

When two chemicals come into contact and react, the reaction often appears to stop when equilibrium is reached. Some portions of the chemicals will have combined into a new chemical product, while other portions appear unaffected. But even in equilibrium, a mixture may remain active: portions of the product may "uncombine" into the reactants while other reactants combine into new product. However, the overall crossover rates balance and there is no net change in the system.

Structural equilibrium is similarly dynamic. A structural element, although at rest, works quietly and unceasingly to resolve the various forces acting on it into an overall force of zero. Without the zeroing of forces, an object will accelerate, decelerate, or change direction.

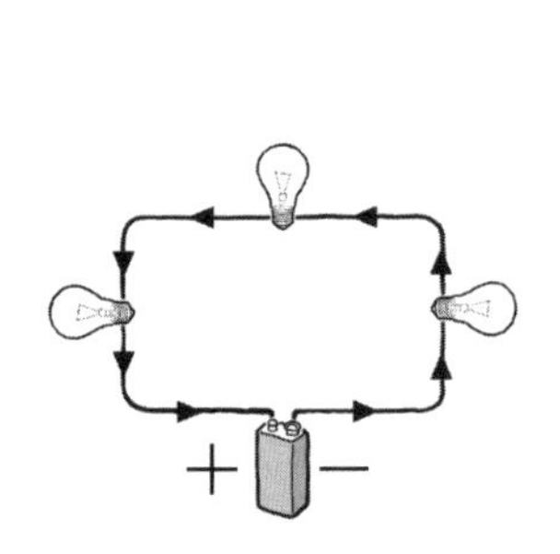

Series circuit

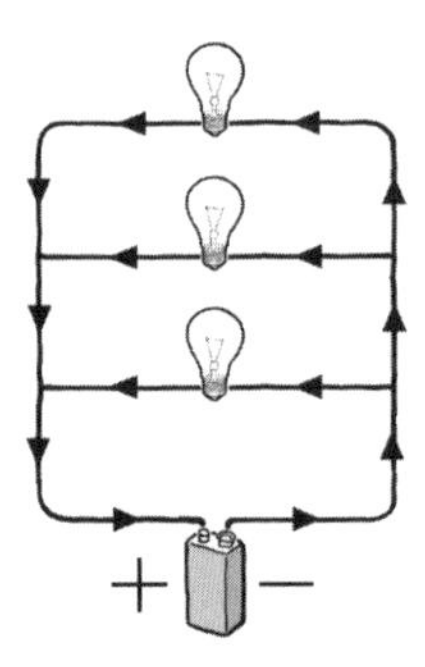

Parallel circuit

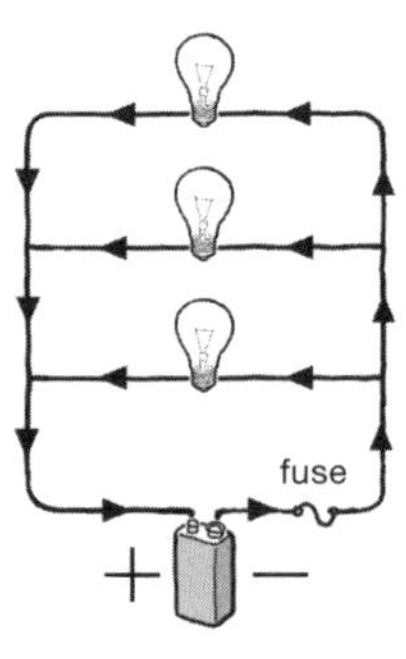

Parallel circuit with
fuse wired in series

An electric current works only if it can return to its source.

A **series electrical circuit** employs a single loop that passes through each device it serves before returning to the source. Each device reduces the voltage available for the other devices; the greater the number of lamps, the more they will all dim. When one bulb burns out, the current path is interrupted and all the bulbs go out.

In a **parallel electrical circuit,** each device receives current from the power source without the current passing through other devices. The voltage is the same everywhere in the circuit, and the number of lamps does not affect their brightness. If one burns out, the others are not affected.

Because of the limitations of series circuits, parallel circuits are used to deliver electricity across cities and within buildings. However, fuses and circuit breakers work by being wired **in series within a parallel circuit.** When a fuse "blows" due to a power surge, it prevents damage to the other devices in the system by interrupting current flow in the manner of a burned-out lightbulb in a series circuit.

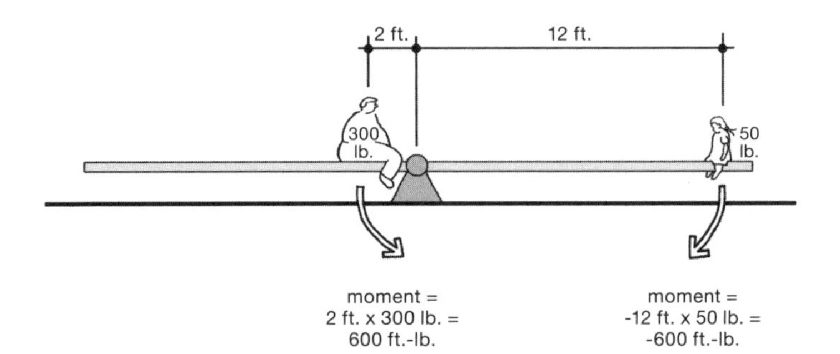
2 ft.
12 ft.
300 lb.
50 lb.
moment =
2 ft. x 300 lb. =
600 ft.-lb.
moment =
-12 ft. x 50 lb. =
-600 ft.-lb.

A seesaw works by balancing moments.

A **moment** is a measurement of rotational tendency around a point, expressed by the equation:

$$\text{moment} = \text{force} \times \text{perpendicular distance}$$

The moment required to rotate a given object around a given point is constant, regardless of where the force is applied. A door, for example, may be opened by applying the necessary force a given distance from the hinge, or by applying twice the force at one-half the distance, or four times the force at one-fourth the distance. Force times distance is the same in all instances.

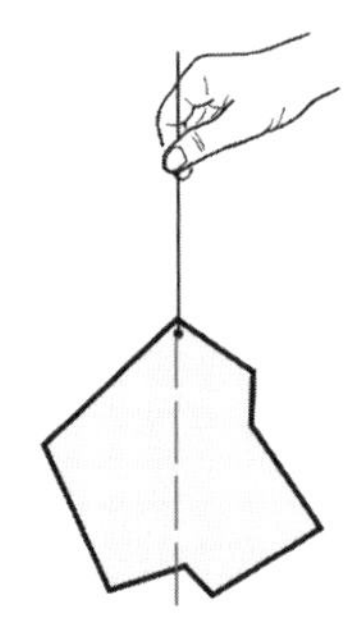

1. Suspend the object from a point and mark a vertical plumb line from it.

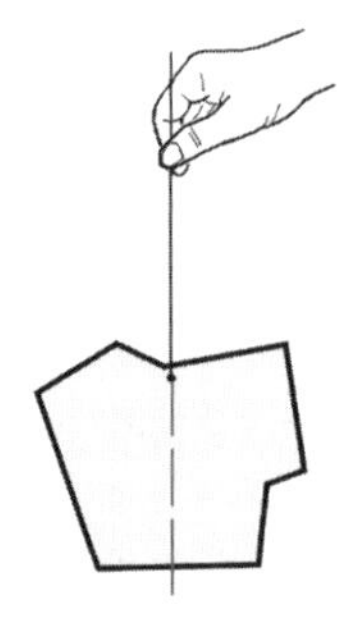

2. Suspend it from a second point and mark a second plumb line.

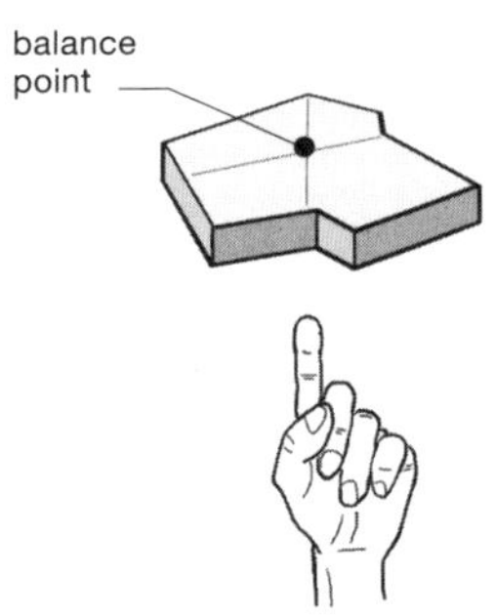

3. The intersection of the plumb lines is the point on which it will balance.

How to locate the center of mass of a flat, irregular object

Center of gravity

The center of gravity of an object is the average position of the particles that comprise it—the point on which it will balance. For objects of uniform density in an environment of uniform gravity, the **centroid** (geometric center), **center of gravity,** and **center of mass** are the same.

86

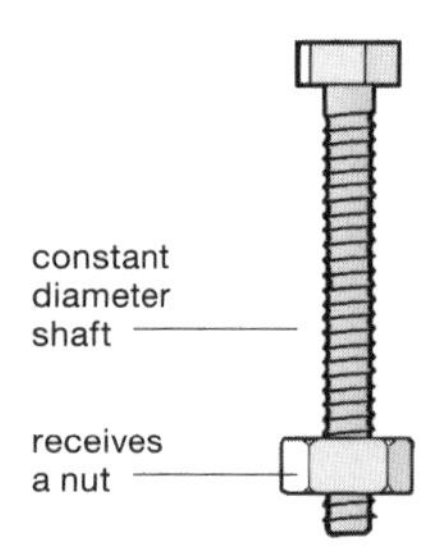
constant
diameter
shaft
receives
a nut

Bolt

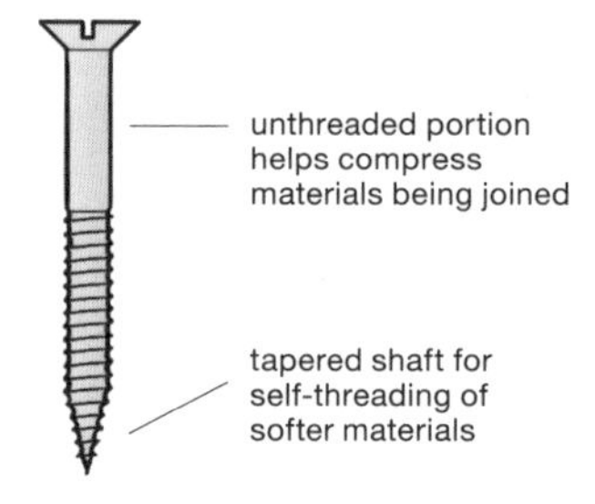
unthreaded portion
helps compress
materials being joined
tapered shaft for
self-threading of
softer materials

Screw

It's a *column,* not a *support column.*

A beam is simply a beam, not a support beam; it's presumed to be there to provide support. Columns and beams are different things: columns are vertical, and beams are horizontal. A column usually is made of steel, concrete, or masonry; if it's made of wood it's a post. There are no "nucular" engineers or "masonary" walls. A building foundation rests on a footing, not a footer. A sink in a kitchen is a sink, while a sink in a bathroom is a lavatory. A bolt and a screw are not the same thing, and by most standards a machine screw is misnamed; it's really more of a bolt. Steel isn't a pure metal, but an alloy; and stainless steel isn't stainless—it just stains less. A hot water heater is simply a water heater; water that's already hot doesn't have to be heated.

87

Articulate the *why,* not just the *what.*

When passing a concept to other designers to develop, make known the reasons for the decisions made to that point, whether technical, ergonomic, personal, or other. By articulating your intent, you will help them understand and preserve the most critical goals of the project while giving them room to investigate possibilities that did not occur to you.

Likewise, when a designer solicits your assistance, ask for an explanation of decisions already made and the goals he or she seeks. This will help head off disappointment if you don't produce the specific result expected.

John Fowler, Kaichi Watanabe, and Benjamin Baker, engineers of the Firth of Forth Rail Bridge, demonstrate its structural system in 1887.

All engineers calculate. Good engineers communicate.

The scientific concepts, analytical processes, and mathematical calculations used to solve engineering problems have been developed over hundreds of years. Early in this period, engineers came to speak a common language, based in mathematics, chemistry, and physics, such that any engineering solution developed anywhere in the world could be read and understood by effectively every engineer.

As more engineering specialties have emerged, engineering "dialects" have arisen. Engineers now must be aware of highly specialized terminology and concepts within their discipline, and as well, must be ever more able to translate them into everyday language that can be understood by clients, end users, and other engineers.

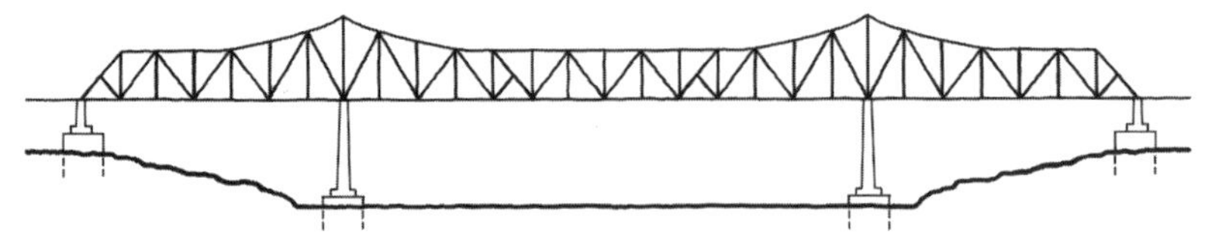

Common double cantilever truss bridge

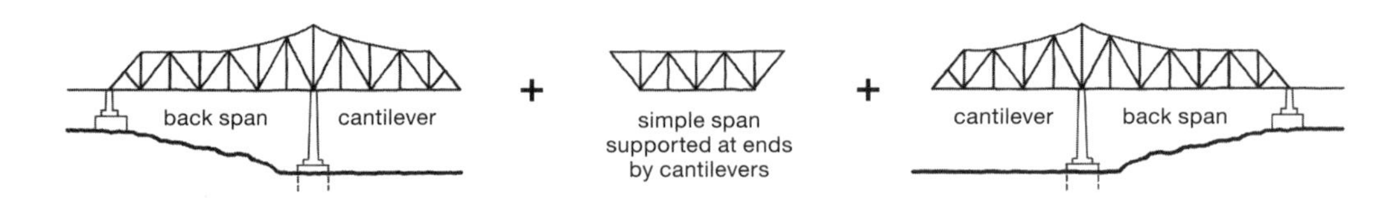

How it works structurally

How to read, but not necessarily name, a cantilever bridge

Most large bridges—cable stay, suspension, steel truss, and some concrete arch bridges—are built by the **cantilever method.** Concrete piers (and often towers) are constructed in a river or chasm, and structural members are gradually extended (cantilevered) in opposite directions from each. This keeps loads balanced during construction, and allows work to proceed from built areas toward unrealized portions of the span. Eventually, the cantilevers from the two piers/towers meet each other in the middle, and the shores at opposite ends.

When completed, cable stay, suspension, and concrete arch bridges do not function as cantilever systems and are instead known by their respective labels. But when a steel truss bridge is built by the cantilever method, it usually functions as a cantilever structure and is permanently classified as such. But oddly, even those steel truss bridges built by the cantilever method that do *not* behave as cantilever structures are permanently classified as cantilever bridges.

painter
musician
technical writer
author
fashion designer
English teacher
computer programmer
clergy
chemistry teacher
engineer
counselor
banker
machinist
merchant

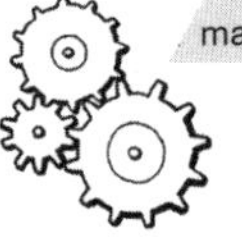

Random hypothesis #2

There are three kinds of people in the world: language people, people people, and object people. Language people find meaningful connections with the world through written, spoken, and symbolic communication. People people seek empathic connections to other people and to human causes. Object people experience the world primarily through relationships with physical things. But object people don't merely "like" objects; they understand the world from the viewpoint of the objects with which they are concerned.

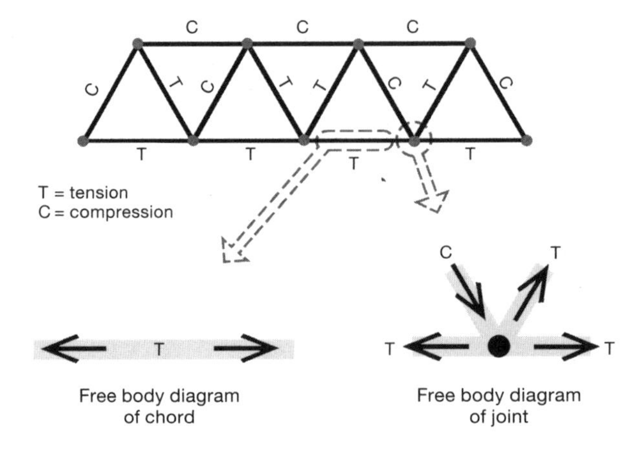
C
C
C
C
T
C
T
T
C
T
C
T
T
T
T
T = tension
C = compression
C
T
T
T
T
Free body diagram
of chord
Free body diagram
of joint

Now you're the chord. Next you'll be the joint.

When struggling to analyze a complex problem, shift your point of view from that of outside observer to that of the thing you are analyzing. If you were that thing, what forces would you feel? What internal stresses would you experience? How would you have to react in order to remain stable and not twist, turn, deform, be pushed over, or be caused to accelerate?

Structural analysis of trusses requires that one continually shift point of view among its chords and joints in this way. If you don't, you can get the force arrows pointing the wrong way. When a chord is in tension, for example, the force arrows point outward from it as one might expect. But at the joint to which the same chord connects, the force arrows point the opposite way—back toward the chord. This is because tension pulls on *all* the components involved, and on each portion of each component. Each component and portion experiences the tension from its own point of view.

$$\text{satisfaction} = \frac{\text{reward}}{\text{input}}$$

Activity satisfaction

$$\frac{\text{reward}}{\text{input}} \longleftrightarrow \frac{\text{reward}}{\text{input}}$$

Activity comparison

$$\frac{\text{reward}}{\text{input}} \longleftrightarrow \frac{\text{peer's reward}}{\text{peer's input}}$$

Situational comparison

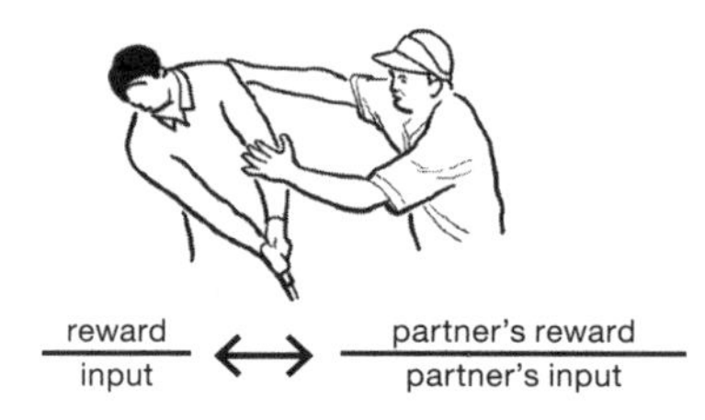

$$\frac{\text{reward}}{\text{input}} \longleftrightarrow \frac{\text{partner's reward}}{\text{partner's input}}$$

Relationship comparison

The engineering of satisfaction

The satisfaction one derives from an activity can be conceptually quantified as the ratio of reward to input. Reward does not have to exceed input for a person to deem an endeavor worthwhile. Rather, a favorable comparison of one's situation to other possible situations or to a peer's situation is the crucial determinant. When people feel fairly rewarded—when their ratio is at least as high as a peer's ratio—they are more likely to be motivated. When people feel unfairly rewarded, they are inclined toward disinterest, demotivation, and resentment. When over-rewarded, they may experience guilt.

In interpersonal relationships, partners might not give equally or receive equally, but their satisfaction will likely be mutual if they experience the same ratio of giving to receiving.

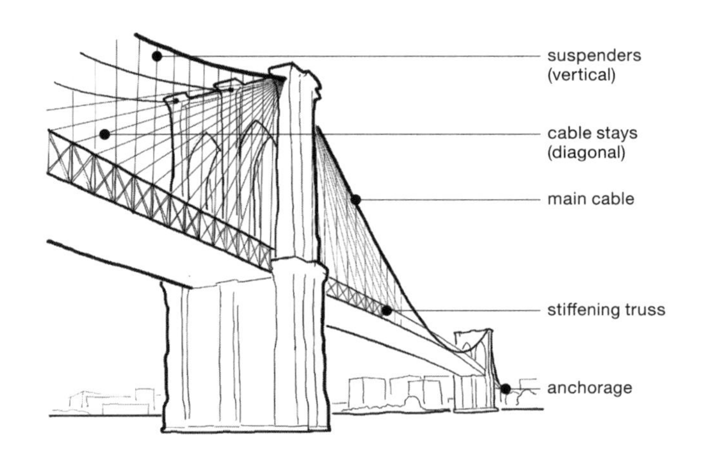
suspenders
(vertical)
cable stays
(diagonal)
main cable
stiffening truss
anchorage

Engineering events are human events.

A beloved symbol of American ingenuity and optimism, the Brooklyn Bridge had an inauspicious beginning. Engineer John Augustus Roebling's foot was crushed by a ferry while he was surveying the location. After having several toes removed and while dying from tetanus, he appointed his 32-year-old son, Washington, to take over the project. Less than 3 years later, the younger Roebling exited a pressurized excavation chamber too quickly and fell ill. Nearly paralyzed, he was confined to his apartment for the next 11 years. His wife, Emily, supervised the work to completion.

In May 1883, after the death of more than two dozen workers, the world's longest suspension bridge, and the first to use steel wire, opened to the public. Washington Roebling could not attend the ceremony. Emily crossed first, followed by 1,800 vehicles and 150,000 pedestrians.

Some New Yorkers were suspicious of the bridge's strength. Not only was it 50% longer than any previous suspension bridge, it had been discovered during construction that the cabling contractor provided inferior wires. At the time, Roebling installed 250 additional cables running diagonally from the towers to the deck. This turned the bridge into a hybrid suspension/cable-stayed structure and gave it its uniquely elegant, cobwebby look. But concerns lingered. Several days after the opening, a woman on the bridge screamed, sending hundreds into a panicked stampede. Twelve were killed.

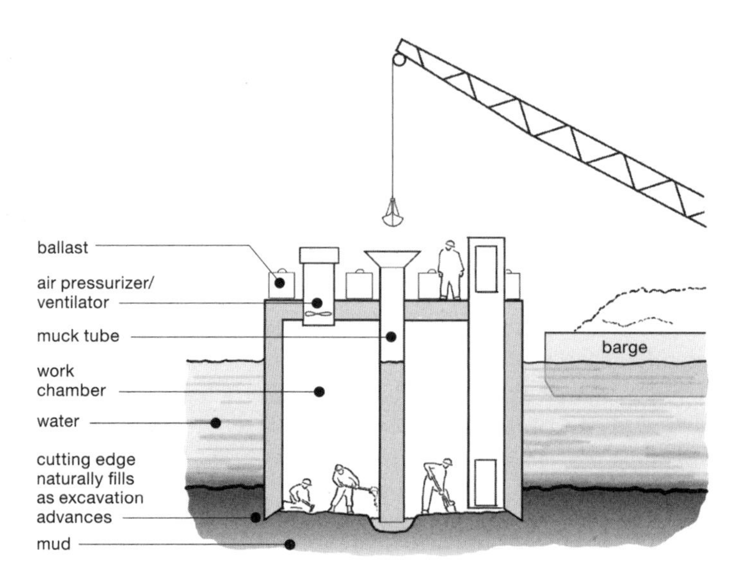

Bridge caisson cross section

There's design besides the design.

A well-designed product *isn't* well designed if the process needed to manufacture it is unrealistic or uneconomical. A brilliantly conceived alternative-fuel vehicle will not succeed without the design and implementation of a refueling infrastructure over a large geographic area. A cleverly resolved construction detail isn't clever if it doesn't leave room for a construction worker to manipulate the tools needed to construct it. A bridge pier that's well engineered won't be built without also engineering a process to excavate earth and pour concrete in the middle of a river.

95

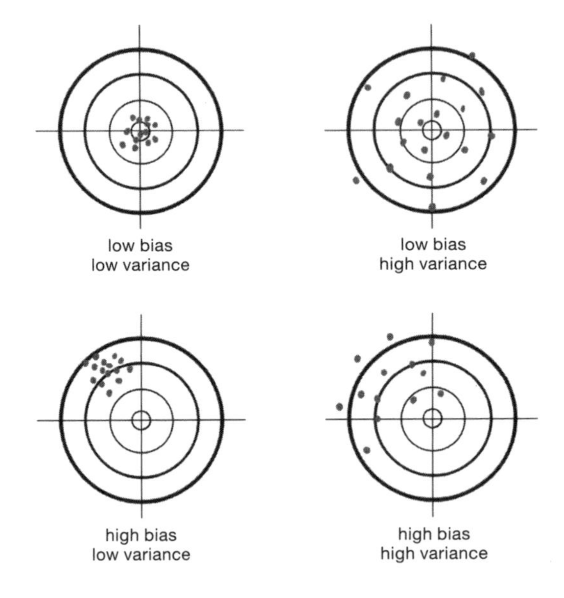

Bias is the difference between a predicted value and the actual value. **Variance** is the average distance between a set of data points and their mean value.

Identify a benchmark against which outcomes will be measured.

An engineering solution must demonstrate objectively measurable improvement. This requires that one establish a starting state against which to compare it. At the beginning of the design process—and particularly when the outcome will be perceived and measured differently by different stakeholders—find agreement on and make known any benchmark that will be used to determine improvement. Take accurate measurements before and after, and revisit the benchmark throughout the design process to make sure it remains relevant. If it no longer is, abandon it, but don't abandon having a benchmark. Identify another, more relevant one that will show you did your job.

96

PRIORITY

“The most important thing is to keep the most important thing the most important thing.”

—DONALD P. CODUTO, *Foundation Design*

97

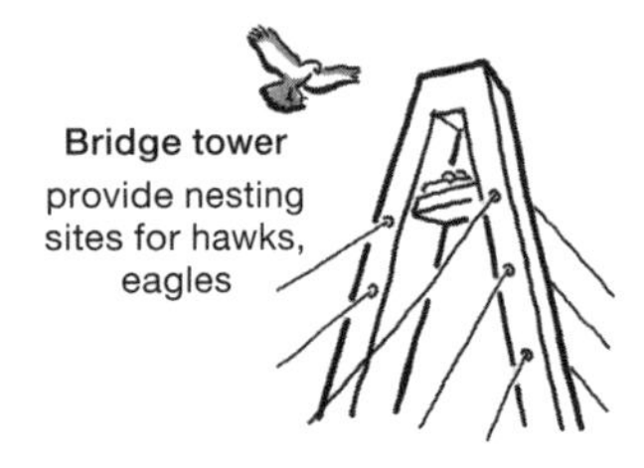

Bridge tower
provide nesting sites for hawks, eagles

Highway barrier
gabion wall construction uses scrap rock, encourages climbing vegetation, discourages graffiti

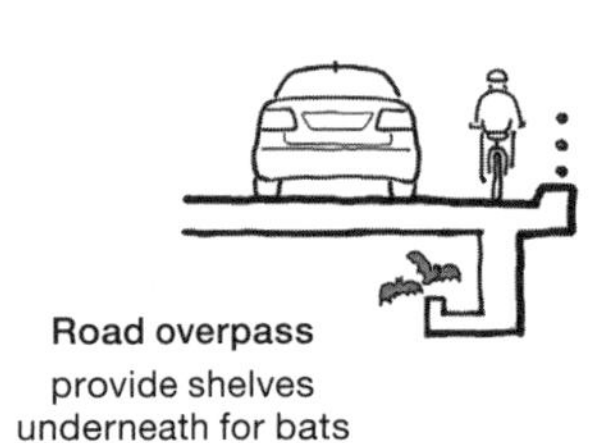

Road overpass
provide shelves underneath for bats

Wastewater treatment
methane off-gas used to power operations

While getting the one thing right, do more than one thing.

Engineering is a field of specialties, and engineers are called on to solve specific problems. Don't get distracted by all the other possibilities such that you forget to do the one thing you must do. But don't become so focused on the one thing that you don't do as much as you can.

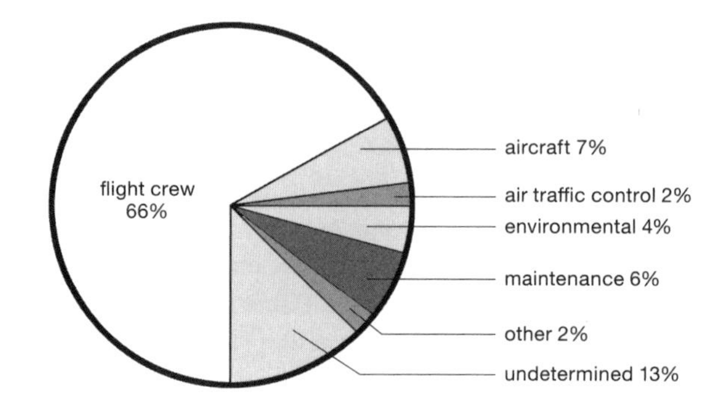

Worldwide causes of airplane accidents

The fix for an apparent engineering problem might not be an engineering fix.

On March 27, 1977, a Pan Am 747 was taxiing on the runway of Tenerife North Airport in the Canary Islands as a KLM Airlines 747 attempted to take off. The resulting collision killed 583 people, the most ever in an aviation accident. Numerous physical contributors were identified, including:

- unusually high traffic due to the temporary closing of a larger airport nearby
- many airplanes parked on the taxiway, complicating taxiing patterns
- dense fog that greatly limited visibility
- no ground radar; controllers had to rely on radio to identify plane positions
- simultaneous radio transmissions that canceled each other out, causing messages to be unheard or misheard, and leading to unauthorized takeoff by the KLM captain despite concern from the copilot

Among the industry-wide changes after the disaster:

- The word "takeoff" was forbidden except when the control tower authorizes an aircraft to take off. At all other times, "departure" or another term is to be used.
- Flight crew across the industry were retrained, with lower-ranking crew encouraged to challenge captains with any concerns, and captains required to consider crew concerns in making all decisions.

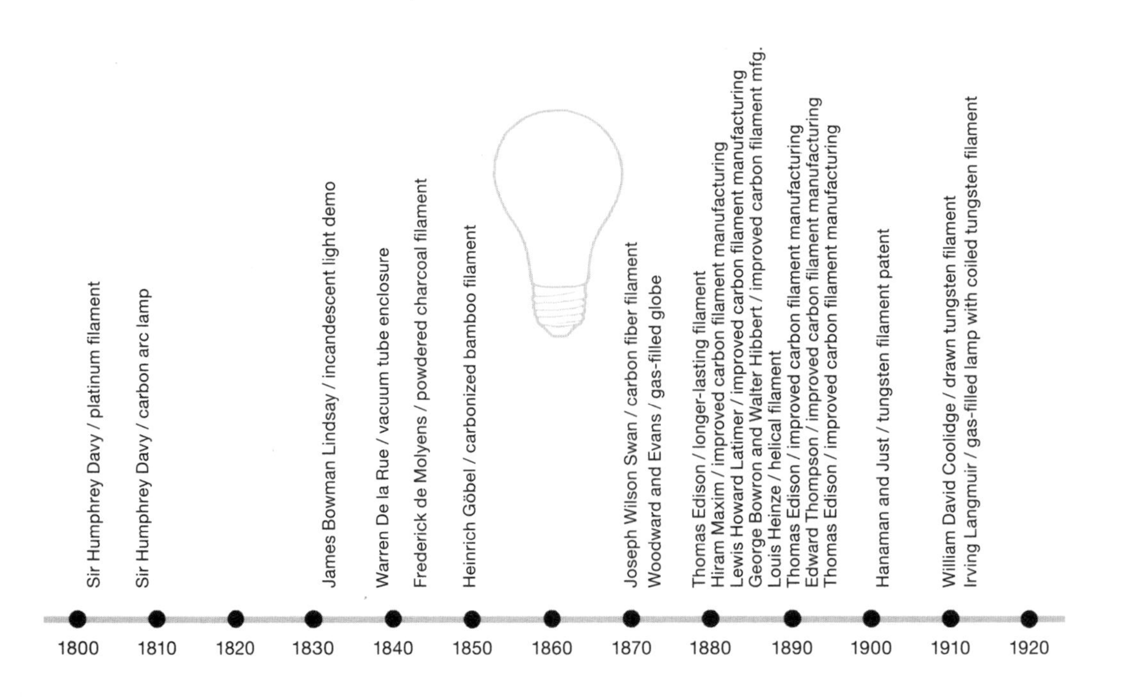

Thomas Edison's help in "inventing" the light bulb

Engineering usually isn't inventing the wheel; it's improving the wheel.

Great inventions usually are incremental steps forward from previous achievements. Gideon Sundback invented the modern zipper in 1917. However, Whitcomb Judson had developed the "Clasp Locker" 18 years earlier and Elias Howe, inventor of the sewing machine, had patented an "Automatic, Continuous Clothing Closure" more than 40 years before that. Sundback replaced the hook-and-eye fasteners of the earlier devices with "scoop-dimpled" teeth, increased the number of fasteners per inch, and provided the familiar slide mechanism that opens and closes the system. Six years after Sundback patented his "Separable Fastener," B. F. Goodrich coined the onomatopoeic "Zipper" when his company introduced galoshes with Sundback's device.

100

Philosopher
contemplates paradigms, meaning, value of human endeavor

Scientist
identifies principles of nature through hypothesis and experiment

Engineer
designs useful things based on proven scientific principles

Technician
inspects, troubleshoots, and implements using known methods

User
desires seamless application; usually has little technical knowledge

The Great Continuum

Engineering is undertaken within a continuum that connects profound human questions to ordinary activities. Engineers who work without awareness of the continuum will be inclined toward performing rote procedures. Those working in awareness of it will be better positioned to adapt to changing times, unexpected challenges, and unfamiliar circumstances. Those working across the continuum may be most apt to contribute something new.

Notes

Lesson 1: Illustration adapted from Mark Holtzapple, W. Reece, *Foundations of Engineering* (McGraw-Hill Science/Engineering/Math, 2nd ed., 2002), p. 9.

Lesson 6: Illustration with regard to Ralph Caplan, *By Design: Why There Are No Locks on the Bathroom Doors in the Hotel Louis XIV and Other Object Lessons* (St. Martin's Press, 1982).

Lesson 52: Illustration adapted from Frederick Gould, *Managing the Construction Process* (Prentice Hall, 4th ed., 2012), p. 64.

Lesson 53: Gary T. Schwartz, "The Myth of the Ford Pinto Case," *Rutgers Law Review*, vol. 43, p. 1029.

Lesson 67: Illustration adapted from John Elkington, *Cannibals with Forks: The Triple Bottom Line of 21st Century Business* (Capstone Publishing, 1999).

Lesson 75: "The Possibility of Life in Other Worlds" by Sir Robert Ball, *Scientific American Supplement* no. 992, January 5, 1895, pp. 15859–61.

Lesson 76: After Jane Jacobs, *The Death and Life of Great American Cities* (Random House, 1961), pp. 430–31.

Lesson 99: Illustration data adapted from "Global Fatal Accident Review, 1997–2006," UK Civil Aviation Authority.

Index

accuracy, 23, 25, 26
aerodynamics, 16, 19, 20
air conditioner, 62
aircraft
  accident causes, 99
  aerodynamics, 19
  reliability, 50
analog recording, 25
arch, 29, 39, 90
Archimedes' principle, 18
automobiles, 20
  aerodynamics, 19, 20
  collisions, 21
  Ford Pinto, 53
  safety, 21, 53

Baker, Benjamin, 89
Ball, Sir Robert, 75
beam
  back span, 12, 37
  cantilever, 12, 32, 37
  castellated, 37, 38
  concrete, 12
  efficiency, 36, 37, 38
  I-beam, 36, 38
  load transfer, 34
  as span type, 29
  tension and compression
    in, 27, 36, 74
benchmark, 96
Bessemer process, 45
bias, vs. variance, 96
black box model, 2
bridges, 90
  Brooklyn, 94
  cable stay, 75, 90, 94
  caisson, 95
  cantilever, 90
  concrete arch, 90
  Firth of Forth Rail, 89
  London Millennium
    Footbridge, 15
  nesting areas for wild-
    life, 98
  nomenclature, 90
  optimal spans, 75
  span types, 29
  suspension, 16, 75, 90, 94
  Tacoma Narrows
    ("Galloping Gertie"),
    16
  truss, 16, 75, 90, 94
building
  code, 41
  exterior wall, 40
  foundation, 18, 30, 31, 34
  lateral loads, 32, 33
  skyscrapers, 32
  structural system, 30, 31,
    32, 33, 34, 35
  vertical load transfer in,
    30, 31, 34,35
buoyancy, 18

cantilever, 12, 29, 32, 74, 90
Caplan, Ralph, 73
center of gravity, 86
Chapman, Colin, 20
chemicals, 81, 82, 83
communication, 87, 88, 89,
  91, 99
compression, 8, 9, 11, 36,
  74, 92
concrete
  vs. cement, 44
  curing, 43
  ingredients, 44
  Roman Empire, 45
  steel-reinforced beam, 12
  strength, 27, 43
cost, 48, 49, 52
crack propagation, 77

design process, *see* problem
  solving
digital recording, 25

Eames, Charles, 73
earthquakes, 15, 32, 33, 55
Edison, Thomas, 100
electricity, 13, 56, 84
electrolytic interaction, 13, 14
energy
  geothermal, 64
  solar, 65
engineering continuum, 101
engineering failure
  black box, 2
  corrosion, 12, 14
  earthquakes, 15, 55
  Kansas City Hyatt walkway collapse, 35
  mean time before failure, 51
  moral aspects of, 47, 53
  natural resonance, 15, 16, 17
  rate of, 50, 51
  safety margins, 27
  structural, 34, 35
  Tacoma Narrows Bridge, 16
  Tenerife North Airport disaster, 99
  unintended load path, 34
engineering family tree, 1
engineering specialties, 1, 2
environmental engineering, 1, 64, 65, 66, 67
equilibrium, 78, 83
error, 23, 47

feedback loops, 78
fluids, 61
force, *see* load
Ford Pinto, 53
Fowler, John, 89
fracture, in materials, 9, 10, 11
free body diagram, 5, 83, 92
friction, 7, 22

gabion wall, 98
galvanic series, 13
Goodrich, B. F., 100

Hammurabi, Code of, 41
heat, 62, 63, 64, 65
Hoare, Tony, 7
Hotel Louis XIV (Quebec), 6
Howe, Elias, 100

invention, 38, 100

Jacobs, Jane, 76
Judson, Whitcomb, 100

Kansas City (Missouri) Hyatt walkway collapse, 35
Kettering, Charles, 38

load, 31
  on beam, 36
  dead and live, 31
  defined, 8
  general effect on bodies, 8, 9, 10, 11
  gravity (vertical), 32
  lateral (wind and earthquake), 8, 31, 32
  multiple vs. point, 37
  path, 34
  safety margins, 27
  transfer to earth, 30, 31, 34

manufacturing stages, 46
masonry, 39
materials
  combining, 12
  effect of environmental factors, 12
  efficiency, 37, 38
  general structural properties, 10, 11, 12, 39
  hardness, 14, 17
  historical development, 45
  longevity, 14
  quality variability, 27
  strength, 10, 11, 12, 27, 39
moment (rotation), 85
moral responsibility of engineers, 47, 53, 66, 67

people
  engineering continuum, 101
  motivation/satisfaction, 93

role in engineering events, 94, 99
three kinds, 91

mean time before failure, 51
measurement, 23, 25, 26, 96
metals, 13, 14
modulus of elasticity, 11
multi-functionality, 54

natural (resonant) frequency, 15, 16, 17
neutral axis, 36

Panama Canal, 60
phasing, *see* scheduling
precision, vs. accuracy, 23
problem solving (and design process), 3, 5, 6, 23, 24, 30, 48, 71, 72, 73, 74, 75, 79, 88, 89, 92, 95, 96, 98, 99, 100, 101

quality, 47, 50, 52
quantification, *see* measurement

reliability, 50
Rockwell hardness test, 14
Roebling, John Augustus, et al., 94
Roman Empire, 1, 45

safety
airline, 99
automobile, 21, 53
chemicals, 81
in environmental engineering, 67
margins, 35
in structural design, 27
system protection, 55
scheduling, 48, 49
service life, 51
site engineering, 57, 58, 59, 60, 66
soil bearing capacity, 31
span
back, 12, 37
optimal length, 75
types, 29
specifications, 42
steel, 12, 13, 14, 27, 35, 36, 37, 38, 45, 75, 87, 90, 94
strain and stress, 8, 10, 11
structural engineering
analysis, 30, 34, 92
failure, 34, 35
safety margins in, 27
stability, 28, 33, 39
Sundback, Gideon, 100
sustainability, 64, 65, 66, 67, 68, 69, 70
systems
analysis of, 74, 76, 92
behavior, 72, 73, 74, 75, 76
deterministic vs. stochastic, 76
scaling of, 75
systematic thinking, 72

Tenerife North Airport disaster, 99
tension, 8, 10, 11, 29, 36, 74, 92
traffic intersections, 21
triangulation, structural, 28, 33
trusses, 16, 29, 37, 92
Tunkhannock Viaduct (Nicholson, PA), 80

variance, vs. bias, 96
vectors, 4, 5, 83

Watanabe, Kaichi, 89
water
global cycle, 68
groundwater, 18
recycling, 69
stormwater management, 66
waste treatment, 67, 69, 70, 98
wood, 27, 40, 45

John Kuprenas is a registered engineer and LEED professional. He lectures in civil engineering at USC and Cal State Long Beach, and is senior vice president and deputy director of the Construction Management Division at STV Group, Inc. His writings have been published in numerous journals and in *The Story of Managing Projects* (Praeger Publishers).

Matthew Frederick is an architect, urban designer, instructor of design and writing, and the creator of the acclaimed 101 Things I Learned series. He lives in New York's Hudson Valley.

www.101ThingsILearned.com